# A Walk Through Time

## From Stardust to Us

*To Jim Raddatz —
Thanks for your contribution on the great team That introduced the first "Walk..." at HP Labs, 1997.
Fred Lieber
Oct 1998*

## The Evolution of Life on Earth

Sidney Liebes, Elisabet Sahtouris, & Brian Swimme

with the Hewlett-Packard "Walk" exhibit text by Lois Brynes

John Wiley & Sons, Inc.

New York  Chichester  Weinheim  Brisbane  Singapore  Toronto

WE DEDICATE THIS BOOK
TO THE FUTURE OF ALL LIFE

All royalties accruing to the Foundation for Global Community from sales of this book will be applied to furthering the "Walk Through Time" project.

Published by John Wiley & Sons, Inc.
Published simultaneously in Canada.

Produced by The Book Laboratory, Inc., Mill Valley, California.
Designed by i4 Design, Sausalito, California.

This publication is designed to provide accurate and authoritative information in regard to the subject matter covered. It is sold with the understanding that the publisher is not engaged in rendering professional services. If professional advice or other expert assistance is required, the services of a competent professional person should be sought.

ISBN 0-471-31700-4

Printed in the United States of America.

10 9 8 7 6 5 4 3 2 1

Library of Congress Cataloging-in-Publication Data

Liebes, Sidney.
    A walk through time : from stardust to us / Sidney Liebes, Elisabet Sahtouris, and Brian Swimme.
        p.   cm.
    "With text by Lois Brynes from the original Hewlett-Packard Walk thru Time exhibit."
    Includes index.
    ISBN 0-471-31700-4 (hardcover : alk. paper)
    1. Evolution (Biology)  2. Life—Origin.  3. Geobiology.
I. Sahtouris, Elisabet.   II. Swimme, Brian.   III. Title.
QH367.L525   1998
576.8—dc21                                                    98-34495
                                                                  CIP

# CONTENTS

# HOW TO USE THIS BOOK

Imagine taking a walk where every foot transports you a million years in time! *A Walk Through Time: From Stardust to Us* takes you on just such a journey. In addition to the informative main text written by mathematical cosmologist Brian Swimme and by evolutionary biologist Elisabet Sahtouris, which tells the lively story of the history of the evolution of life, the book contains the original text and illustrations of the acclaimed "Walk Through Time . . . from stardust to us" exhibition, on tour around the world. This innovative transportable exhibition tells the story of life's evolution via an illustrated one mile-long walk highlighting major developments. In the book, the exhibition text runs along the bottoms of pages, accompanied by the illustrations from the exhibition, and a time-line showing position along the "Walk," in MYA (millions of years ago). Each foot in the actual exhibition represents one million years. At this scale . . .

| 1 foot | = | typical time for continents to move and earthquake faults to slip 20 miles |
|---|---|---|
| 0.1 inch | = | time span back to the end of the last Ice Age |
| 0.001 inch | = | human lifetime |

This book offers a dual experience that allows you both to savor the astonishing story of life's evolution through the lively narrative of the main text and to enjoy the exhibition in the comfort of your own home.

Comprehensive information regarding the exhibition, including bringing the "Walk" to your community or organization and instructions for using the book to convert a favorite one-mile stroll into a self-guided "Walk Through Time" are available from the following sources:

World Wide Web:
http://www.globalcommunity.org/wtt

Foundation for Global Community
222 High Street
Palo Alto, CA 94301-1040
U.S.A.
Tel: (650) 328-7756
Fax: (650) 328-7785
e-mail: wtt@globalcommunity.org
http://globalcommunity.org

# PREFACE

*Each of us carries within us a worldview,
a set of assumptions about how the world works
—what some call a paradigm—that forms the
very questions we allow ourselves to ask and
determines our view of future possibilities.*

Francis Moore Lappe, *Rediscovering American Values: A Dialogue
That Explores Our Fundamental Beliefs and How They Offer Hope
for America* (Ballantine Books, 1989)

There are many views of the origin of life, and
diverse paths to experiencing awe, wonder, and
gratitude for the universe and the gift of life.
This book offers the lay reader a scientific understanding of the evolution of life on Earth, in the
conviction that the evolutionary story constitutes
a rich and deeply meaningful context for identifying and addressing the most critical issues of
the future. Regardless of one's beliefs regarding
origins, the future of life on Earth will be substantially determined by the extent to which
humanity unites within the next few years in
commitment to the future.

The title of this book is drawn from an exhibition created by employees of HP Laboratories,
the central research labs of the Hewlett-Packard
Company, in collaboration with external experts,
for an HP Labs 1997 Earth Day "Celebration of
Creativity." Consisting of approximately ninety
large panels of text and colorful illustrations

depicting significant events and themes in the
evolution of life on Earth, this mile-long, transportable "Walk Through Time . . . from stardust
to us" unfolds a scientific understanding of the
five-billion-year evolution of Earth and life upon
it. Each foot along the "Walk" represents one
million years. At this scale, microbial life appears
four thousand feet in the past and remains the
sole life-form for the better part of the mile. For
each foot along the walk, continents typically
drift and earthquake faults slip twenty miles;
humans appear three feet ago (and a single
human lifetime spans one thousandth of an inch);
the last ice age ended a tenth of an inch in the
past; and the world's human population doubles
in size each half a thousandth of an inch. The Sun
can support life on Earth for another two to
three thousand feet.

This book, which contains the text and
illustrations of the physical "Walk" plus a story
integrating the content of the panels, is designed
both to be enjoyed as a traditional book and to
enable a self-guided, private "Walk Through Time."

The "Walk" had its genesis back in the 1960s,
when I lectured to Stanford University students
on environmental issues. I used to line the walls
of the lecture hall with a hundred-foot-long
mini-"Walk" terminating in a twenty-foot-high
"population plot." I noted that the plot doubled

5

in height each horizontal millionth of an inch. I felt that science could contribute powerfully to our perspective regarding our place in the scheme of things: an appreciation of the enormous span of time over which life has evolved on Earth; an awareness that the sun could support life on Earth for another couple of billion years; knowledge that humanity was precipitating what could rapidly become the greatest extinction of species in the past 65 million years; and recognition that human actions today threatened the diversity, quality, and stability of life on Earth for tens of millions of years into the future. I believed that if people knew they would care.

On the first Earth Day, in 1970, I proposed a three-mile-long "Walk" unfolding the 15-billion-year story of the evolution of the universe, Earth, and life, from the Big Bang to the present. One of my weaknesses is that I always expect others to share my enthusiasms. I expected scores of volunteers to pour out of the woodwork. It turned out that no one was crazy enough to want to help, and I wasn't crazy enough to go it alone. Had we proceeded, it would not have been possible to have provided anything approaching the richness of content within the present "Walk," as few of the relevant exquisite discoveries of modern science had been made at that time. I set the "Walk" aside, as it turned out, for a quarter of a century.

I retired in 1997 after fifteen years at HP Labs. When I announced to my colleagues in 1996 that I planned to retire, and they asked why, I said that, besides being sixty-six years old, I wanted to create a "Walk Through Time," which had been on my mind for twenty-five years. I told them that I planned to raise private funds to build the one-mile "Walk" (scaled back to span from the formation of Earth and life upon it to the present). One person I spoke with was my friend Barbara Waugh, Worldwide Personnel Manager for HP Labs. Barbara, who vigorously strives to "raise the bar on corporate citizenship," and fans the flames of the passionate, was saddened at the suggestion that I had to leave HP to, as she put it, "fulfill my dreams." She is well versed in the company's core values, one of which, articulated by Dave Packard in 1958, declares that "The Hewlett-Packard Company should be managed first and foremost to make a contribution to society." She suggested "HP might need your dream to address its future." Laurie Mittelstadt, another close mutual colleague, immediately volunteered to help, and Bill Shreve, director of my lab, freed me of all other responsibilities to work on the "Walk." We talked with the HP Labs' senior management. The "Walk" won favor as a powerful context for identifying and addressing fundamental issues of the

6

future, whether social, personal, corporate, or governmental. We received enthusiastic support from both Joel Birnbaum, HP Senior Vice President for Research and Development, and Lew Platt, CEO, President and Chairman of the Board of Hewlett-Packard.

HP committed to piloting the "Walk" for the 1997 Earth Day "Celebration of Creativity," with the intent of subsequently donating the "Walk" to a nonprofit organization for further development and a worldwide presentation. It was gifted by Hewlett-Packard in 1998 to the Foundation for Global Community, an educational nonprofit in Palo Alto, California. The Foundation's "Walk

Through Time" project is staffed by volunteers committed to sharing the "Walk" with the world. As of this writing, the "Walk" has been presented in a dozen venues in three countries, including corporate settings, conferences, museums, and community celebrations.

We hope that the scientific understanding of the story of our origins will become as meaningful for you as it has for us.

Sidney Liebes
Atherton, California
Summer 1998

## COSMIC PROLOGUE

The origin of the universe is an immense fire that will never again appear in the universe's expansion into time. This primal universe shines forth with such titanic radiant energies that we humans evolving out of this blaze can still feel its heat 15 billion years later. It was the 1965 detection of this heat by radio astronomers that convinced scientists that the Big Bang theory was true.

The universe did indeed begin in an explosion of energy powerful enough to send all matter flying apart for billions of years into the future. The primal fire when the universe is one-trillion-trillion-trillionth of a second old is a unique, onetime event. So though we use the word fire with its connotations of a red-orange

8

*Origin of the solar system*

*Bacterial life takes hold.*

| 5,000 MYA | 4,000 MYA | 3,000 MYA |

### INTRODUCTION TO THE WALK
DEEP-TIME TERRAIN *What to Take with You...What to Leave Behind*

The fossil record offers a narrow window through which we try to glimpse vast vistas. Beings with hard parts have favored histories. Dates of evolutionary events are known to varying degrees of accuracy. | Life is a story of permanence and change. Exuberant and innovative, it is also deeply conservative. Traditions and relationships of life today provide clues to the past. DNA is a tangled repository of ancient history; its sequences reveal the presence of the past. Like evolutionary wax tablets, they reveal layers of time and change. | Evolution connotes "change" over time, not "progress." How recently an organism evolved does not define its "worth." While some organisms may be more specialized or complex, all share an equally long evolutionary history. |

flame flickering above burning wood, we need to realize that the primal fire is different. It is a fire that is a billion trillion times hotter than the center of the sun. It is a fire that is a trillion trillion trillion trillion times denser than rock. No one word by itself will ever adequately convey the nature of this beginning, for all of our words have been forged in a very different context—a less violent, less cataclysmic context, but a context nevertheless that came out of the original moment itself. Our words refer to this later world in which we live, but all that is about us descends from that compact beginning and thus each thing and even our words themselves must carry some faint and stable relationship with the original fire. Though we have no single word sufficient for that singular event at the beginning of time, perhaps with a spectrum of words—fire, rock, blaze,

First protists. Eukaryotic cells (cells with nuclei and organelles) evolve through symbiosis.

First animals (marine)

First plants, first fungi

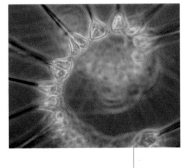
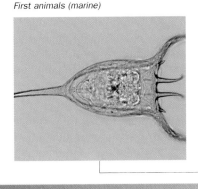
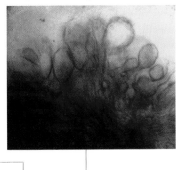

2,000 MYA

1,000 MYA

PRESENT

9

**Evolution is not linear. Organisms and species do not just evolve or become extinct—they also anastomose (fuse together). In profoundly moving ways, life often grows in on itself, bringing previously evolved beings together into new partnerships.**

radiant energy—we can with all humility intimate something of that originating mystery.

The primal brilliance expands briefly and then suddenly, with great fury, enters upon an even more explosive expansion that physicists designate with the phrase inflation, an exponential billowing forth in which the elementary particles, the first material beings, are torn out of a deep well of potentiality and allowed to enter the adventure of evolution. So many exotic things were destined to come forth, but not all at once here at the beginning. This is the time for the emergence of the elementary particles, and their many types bubble forth in a profusion unique to this moment.

Surely the most important insight into the nature of cosmic fecundity concerns time itself. We have been conditioned to regard time in its mechanical and watchlike aspects, almost independent of the universe itself; but a journey through evolutionary time reveals something more. Time is not primarily a clock, nor is time simply an abstract measurement. A meditation on the billions of years of the universe's process provides a glimpse into time as a measure of the universe's creativity. In this understanding, the beginning of the universe is the time of the elementary particles in the sense that it is then and only then that every elementary particle, including the most bizarre, is given a chance at existence.

An infinite concentration of radiant energy

In 1054, an earthling in China records the explosion of a massive supernova 6,300 light years away. The explosion creates the Crab Nebula, which is still expanding 50 million miles per day.

10

## 4600 MILLION YEARS AGO

### PULSE OF THE SPHERES

We begin our "Walk Through Time" with the formation of the Sun and Earth, two-thirds of the way through the history of the universe and 4,600 million years before human beings appear. | About 15,000 million years before humans evolve (two miles behind you, at one million years to the foot), the universe exploded out of the void in a "big bang." Early stars cycled through life and death. Supernova explosions spewed elements cooked in their interiors into space, creating dense clouds of molecules and dust. | New stars and planets form as nebulae contract and condense under the force of their own gravity. In a galaxy we call the Milky Way, the massive center of one such nebula contracts to form our Sun. Orbiting gas and dust accrete (grow by being added to) into

11

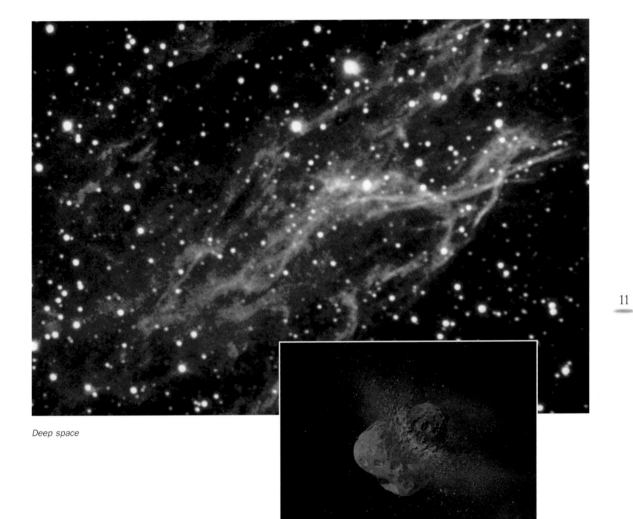

*Deep space*

**planetesimals which then collide to build the planets and moons of our solar system.**

*Crashing planetesimals begin to build Earth.*

enables this vast melange of particles to thunder forth and expand, each one emerging with its companion antiparticle. When such a particle meets an associated antiparticle, they both suffer annihilation into an explosion of light. In the extreme heat of the first moments of the universe, such an annihilation is instantly followed by the birth of yet another pair of particles, perhaps of a different nature, but these two will soon suffer annihilation as well. So long as the universe blazes with such energies, this dance in and out of existence continues and would continue for all time but for a change in the universe as a whole.

As the universe expands, the temperature steadily falls from its initial heights and very quickly the background energy of light is no longer high enough to draw forth particles and antiparticles from the well of potentiality. When the universe is one-millionth of a second old and its temperature has dropped to 10 trillion

degrees, a new threshold has been reached. One by one, each of the protons annihilates with an associated antiproton just as before, but now the temperature is too low to draw forth new protons. As particles and antiparticles continue to disappear, it appears as if the entire future course of time is now set: the universe will continue to expand, the particles will continue to disappear,

*The early Moon, still quite close to Earth, casts a stunning shadow.*

**4500** MILLION YEARS AGO

STAR STUFF OF LIFE

The early planets consist mostly of compounds of heavier elements. As the planets grow, their gravitational fields increase, drawing in nebular dust, planetesimals, and carbon-rich meteorites. Ultimately,

Earth is massive enough and cool enough to retain lighter gaseous compounds of carbon, nitrogen, oxygen, and hydrogen, the star stuff from which life will spring.

12

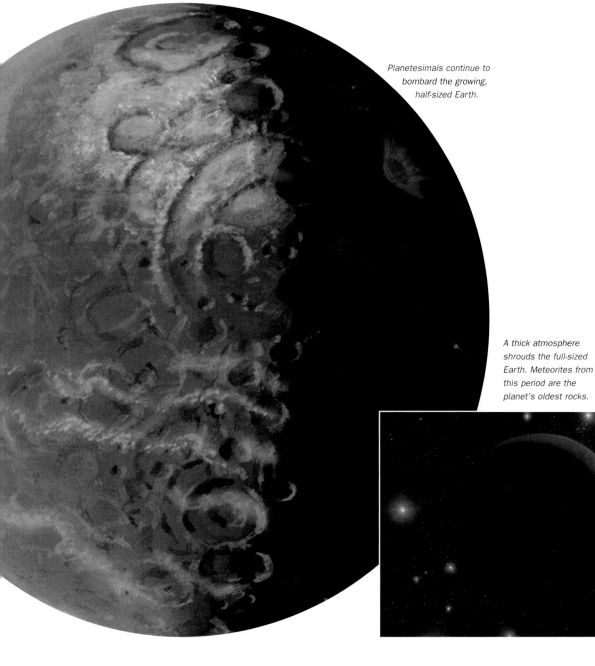

Planetesimals continue to bombard the growing, half-sized Earth.

13

A thick atmosphere shrouds the full-sized Earth. Meteorites from this period are the planet's oldest rocks.

and eventually there will be nothing left but light that will cool and darken as it slowly drops down through trillions of years toward an absolute zero point of temperature.

But the fundamental processes of matter production contain an asymmetry, a slight bias for matter over antimatter. For every 100 million particles of antimatter, there are 101 million particles of matter, and when they meet they all annihilate each other but for a single proton surviving the catastrophe. In a perfectly symmetric universe, nothing but light would be left after such an event. But because of this intrinsic bias, a tiny sliver of the original matter survives this vast conflagration. The universe we have about us today is only one-200-millionth of the original universe. Out of this remnant, the future will be created.

By the end of the first second, the annihilations have finished, leaving a world of stable protons, neutrons, and electrons, all of which now enter upon the work of building up the universe. Irresistibly drawn together by the strong nuclear interaction, the protons and neutrons discover a mutual attraction leading toward higher orders of being, the light nuclei. This process is but one instance of a universal dynamic in cosmic evolution. At the level of elementary particles, this power of allurement that leads to a new system

or community is called quantum chromodynamics; at the level of atoms it is called quantum electrodynamics; at the level of molecules it is called chemical affinity. In each case, a field of attraction initiates a complexification as separate beings are drawn together into composite entities.

The universe is three minutes old, its temperature has dropped to a billion degrees, and the protons and neutrons seal themselves together into atomic nuclei. Previous to this moment the nuclei could not be constructed, for if any had emerged they would have been destroyed instantly by the energetic particles of light, the photons. But three minutes after the cosmic beginning, the conditions are suddenly favorable for nuclei to emerge.

This process of nucleosynthesis brings forth the helium, lithium, and deuterium nuclei and then ceases. Before these first nuclei can join together to build up still larger nuclei, they are swept apart by the cosmic expansion of spacetime. Further nucleosynthesis becomes impossible and thus cosmic construction comes to a halt. The matter of the universe consisting of these atomic nuclei, photons, and various elementary particles continues to expand and cool.

After 300,000 years, the universe arrives at the time for its next spectacular transformation from electrons and nuclei into atoms. The first atoms of hydrogen, helium, and lithium are many millions of times larger than any of the particles or nuclei of the previous era. But even more significantly, the very existence of atoms alters the large-scale dynamics of the cosmos. In the previous era, before the atoms, radiant energy blew matter apart just as easily as a tornado carrying particles of dust; but with the universe moving into its atomic phase, the photons can now pass right through the matter without altering their course.

Matter can now respond to its own mutual attraction and begin following new pathways of allurement. Concentrations of hydrogen and helium soon arise that further amplify and focus the gravitational attraction and the atomic universe soon breaks apart into clouds. Inside these clouds, which are the primal galaxies, hydrogen and helium rush together, heat up, and burst forth as the first stars. What has been a vast expanding mist now shivers forth into the structures of 100 billion galaxies, each one an elegant whirling dance of billions of burning stars. Though the process is a complex one requiring hundreds of millions of years, and though physicists still lack a complete and detailed understanding of how the atoms gathered into galaxies, from the perspective of cosmological time we can picture the emergence of galaxies as similar to the way a cloud of water vapor, under the right conditions, will

suddenly flutter forth into a billion snowflakes.

The universe has a bias for complexity. The universe began as elementary particles, then began constellating into galactic structures. The creativity of the universe accomplishes all this through its own intrinsic self-assembling or self-organizing dynamics. An atom, for instance, is not put together by some agent outside itself. An atom is a group of particles that organizes itself into a whole and coherent system. So too on the larger scales. A galaxy is more than just an aggregation. A galaxy is a self-organizing community of stars. Presumably, these self-organizing dynamics are present in a potential way throughout time, but the universe needs to reach a particular level of complexity before such dynamics can act. After 500 million years a new threshold of organization was reached, and the amorphous clouds of hydrogen and helium began to assemble themselves into elliptical and spiral galaxies, one of which will eventually be called the Milky Way Galaxy, our home.

Each new level of complexity in the universe is accompanied by a new level of vulnerability. The atom is more fragile than the proton. Organic molecules are much more fragile than atoms. And when we come to a system as complex as a star, we find a complexity so vulnerable to destruction that it is only with a constant

15

supply of energy that it is able to maintain its structure. Every star continues to shine only because, moment by moment, it transforms into energy some of the matter in its core. If it were to stop this self-consumption, it would cease to exist as a star, for its matter would implode upon itself. The food of hydrogen is transformed by nuclear fusion into the waste of helium, and in the process the star gets the energy it needs to hold off collapse under its own intense gravitational attraction. When it uses up the plentiful hydrogen, the star compresses itself and heats up further and begins consuming its helium and in the process creates carbon. If even this helium is used up, the star responds by collapsing down further and heating up until it can consume carbon in its quest to hold off gravitational collapse and survive yet another day.

In time, any star will use up all of its elements. Without any more food, its demise is inevitable.

Though it has perhaps burned brilliantly for a billion years, it is now doomed to grow ever dimmer and slough off its luminosity as it sinks into the dark—lost and forgotten in one of the cold chasms of space. But here too, as with the earlier moment involving the annihilation of matter with antimatter, the universe shows a surprise that emerges from its very depths, for not every star will end by slowly winking out.

If the star is of a privileged size, it continues to consume its elements, and then to consume the elements it has constructed, all the way up to iron itself. Now this star, like every other star, will begin its collapse upon itself. No longer obstructed by the radiant energies of the stellar nuclear fusion, the star's matter rushes madly toward its common center of attraction and creates a concentration of heat unlike any moment since the very birth of the universe, a concentration into a single supremely dense point that then blasts

**4400** M I L L I O N   Y E A R S   A G O

SETTLING IN

Radioactive elements inside Earth decay and planetesimals bombard Earth and her sister planets, causing tremendous heating. Heavy metals sink to form cores, while volcanoes spew lighter elements into the atmospheres. This "outgassing" transforms the early atmospheres. | The sister planets of the inner solar system—Mercury, Venus, Earth, Mars— reach their current configurations during this phase.

Venus          Mars

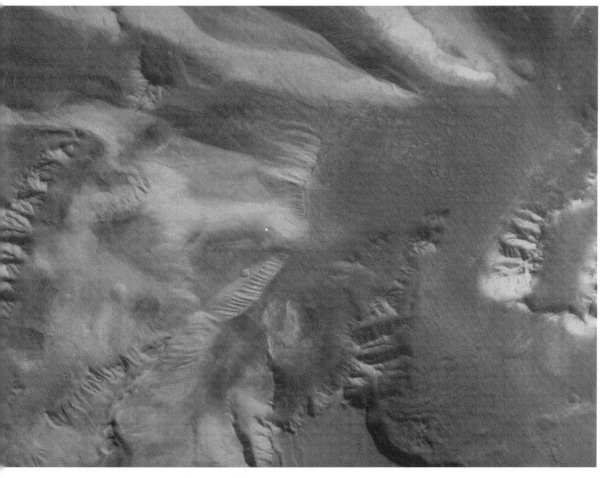

*Mars-scape. We journey into deep time when we journey into space. As we study planetary evolution, we better understand the environmental evolution of Earth and the central role life plays in shaping our home planet.*

apart in every direction in a display of energy that is among the most spectacular in the entire universe. Its end is sheer brilliance, an intensity matching 100 billion stars. And this explosion of the supernova is an end that is, surprisingly, not really an end.

The first supernova explosion utterly destroys the star itself, but in the destruction all the remaining elements of the universe are synthesized for the first time. These amazing new beings are then carried off into the galaxy's processes of creativity, processes that had never proceeded with these elements before. Carbon and nitrogen and potassium and uranium and boron, all of them blasted forth by the stellar winds of the explosion, are then carefully kneaded into the clouds of the galaxy, clouds that have been waiting for this opportunity for billions of years.

Before the first supernova explosion there is no carbon, no nitrogen, no oxygen, no phosphorus, no sulfur anywhere in the galaxy. Suddenly enriched with these cosmic treasures, the Milky Way is able to bring forth entirely new stellar systems. Five billion years ago our galaxy ignites yet another cloud, one seeded with the creativity of such supernovae. As this cloud collapses and evokes its fusion reactions and becomes our Sun, a remnant of the cloud continues spinning around the new star. This remnant breaks into ten bands of matter that cool and accrete into Mercury, Venus, Earth, Mars, the asteroids, Jupiter, Saturn, Uranus, Neptune, and Pluto. The supernova that reached its end has now become a new beginning—a planetary system composed of all the elements and their dazzling possibilities. Rising up into existence, the Sun and Earth together are poised to give birth to a new kind of beauty, one that grows out of all the universe has given birth to thus far.

18

## 4300 MILLION YEARS AGO

ROCK *Not Always a Hard Place*

Earth processes cycle so powerfully that little evidence remains of early planetary faces and flows. Critical and intriguing clues lie in the formulation of Earth rocks and minerals when compared to those on the Moon and other planets. | The oldest mineral crystals on Earth, zircons from western Australia, date back 4,300 million years. The oldest Moon rock brought back by Apollo astronauts is 4,200 million years old. Granite rocks found near Canada's Great Slave Lake go back 3,960 million years. Scientists believe these evolved from even older crustal material, melted and remelted by the restless Earth.

Explorers have not found the rocks that held the early zircons. Perhaps they metamorphosed (changed in form) in the continual cycling of Earth's mantle and crust. Rocks seem like such a "hard place" from our human time perspective. As inflexible as they appear, even rocks change form with heat, pressure, and time. Continual transformation in the geological cycle foreshadows the patterns of life.

19

## THE EVOLVING STORY OF OUR EVOLVING EARTH

Ten billion years of cosmic evolution brings us to the birth of Earth. It takes another five billion years—half again as long—until we humans arrive on the scene and try to understand our awesome origins. Today the concept of the cosmic evolution of Earth as a planet is commonplace—for we can actually see Earth from space, along with other planets, cosmic clouds, and distant galaxies. The evolution of Earth's creatures, from the earliest and tiniest bacteria to ourselves, is also real to us. We encounter it everywhere, from museums full of fossils to computer animations on giant movie screens. It is our story.

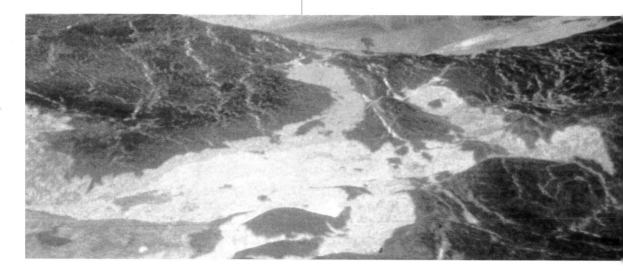

### 4200 MILLION YEARS AGO
### THE RAINS REIGN

As Earth cools, water in the atmosphere condenses: torrential rains fall on, and on, and on. Great seas form. Exuberant volcanoes expel hotly agitated deep earth to the surface. | The hot, early atmosphere dances with an abundance of carbon dioxide, nitrogen, and water, and lesser amounts of methane and ammonia. Intense energy sources available on the primitive Earth form biologically important molecules from these gases. Candidate sources include solar UV radiation (there is as yet no UV-protective ozone shield), radioactivity, and hot oceanic vents.

But our great-great-grandparents, representing only a few human generations in the past, had a very different picture of origins. Before Charles Darwin, people who thought about it at all believed Earth was only six thousand years old, measured in biblical generations from Adam and Eve, however ancient that seemed at the time. Most people were shocked when scientists such as Darwin, late in the nineteenth century, told them that Earth was vastly older, and that they were kin to apes and other animals that had evolved gradually, one kind from another. Imagine their reaction if Darwin had told them they were even kin to bacteria (or what most people refer to as germs). Today we accept Earth's age as billions of years, and most of us can

21

*Earth's geology and biota are a single, tightly coupled evolutionary system. Rocks carry tales as traces of bygone biospheres.*

## CONCEPT NOTE

### GAIA

The Gaia hypothesis, developed by British atmospheric chemist James Lovelock and U.S. biologist Lynn Margulis, proposes that planetary processes are neither simply geological and chemical, nor just geochemical; they are geophysiological. Variables such as the mean global temperature, the gaseous composition of the atmosphere, and the salinity and alkalinity of the oceans are sensitive to biological activity.

contemplate microbial ancestry without batting an eye. But wait: if the cosmos evolves, if Earth and its creatures evolve, and if our human knowledge evolves with the most blinding speed of all, how do we really know that we understand evolution now?

The answer is that our present understanding is far from complete, and not to be taken as dogma. Like the planet itself, our story will continue to evolve. It is virtually impossible to know just how we will understand our evolution in another century.

The observations we make now will not change in essence, but they will certainly be augmented in quantity and precision. Nor is the time frame of the past likely to change substantially. But year by year, we extend our vision with ever more powerful telescopes and microscopes, scanners, counters, and dating techniques, and with explorations of rain forest canopies, ocean depths,

deserts, and polar ice. We continue to see farther into the vast reaches of outer-space macrocosms and deeper into the inner reaches of quantum and molecular microcosms. The more we see, the more the way in which we interpret what we see changes. In short, our current evolution story is itself evolving with breathtaking speed, and will surely continue to evolve for a long time to come.

It is from the leading-edge trends in evolutionary science that we can make our best predictions of how this story will evolve. Four such trends will be evident as the story unfolds in these pages: a systems or ecological perspective, an emphasis on our microbial ancestry, the DNA revolution, and the new understanding of life's creativity in response to crisis.

*The Systems View*

Ecological or systems thinking involves a shift away from tracing the evolution of individual

23

*Nature and art are too sublime to aim at purpose, nor need they, for relationships are everywhere present, and relationships are life.*

Goethe

species' lineage against environmental "backdrops." Increasingly, scientists see evolution systemically and ecologically—as the simultaneous and intertwined co-evolution of all Earth's species. This way of seeing evolution brings it into new focus, resolving environments into ecosystems—complex webs of co-evolving, interdependent species. Each species helps shape every other, and each is shaped by the others.

Even our most basic distinctions between geology and biology—the designated domains of nonlife and life, inanimate and animate—are blurring. We observe that the same atoms over time move from rocks to microbes, to plants, to animals, and then on, with decay, into soils, sediments, and back into rocks. As figure and ground merge, the old view of rabbits in habitats becomes a view of "rhabitats." We amass evidence that Earth's rocky crust (the lithosphere), soils, waters (the hydrosphere), and atmosphere are permeated, altered, produced, and chemically regulated by living creatures (the biosphere), especially microbes.

Carrying this trend further, we find some scientists reviving an old, even ancient, concept common to most of Earth's human cultures —that Earth itself is alive. Most scientists today recognize that Earth's lithosphere, hydrosphere, atmosphere, and biosphere are dynamic systems—self-organizing and inseparably interconnected. Yet the concept of a living Earth remains controversial. How the controversy is resolved will depend largely on how we agree to define life—a matter itself still unresolved.

*Our Microbial Ancestry*

Another major trend is inspired by the new recognition that multicelled creatures such as

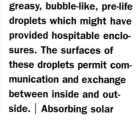

24

---

| 4100 | MILLION YEARS AGO |
| --- | --- |
| PRIMORDIAL SOUP | |

The stage is set for life, but how does the play open? | Seeking to answer this question, some scientists investigate liquid and atmospheric chemical combinations that, when sparked, spontaneously generate the components of all life. Others investigate RNA and DNA, key information molecules that make up genes. Yet others explore the potential for life to begin at deep-sea vents, or in "bubbles," meaning greasy, bubble-like, pre-life droplets which might have provided hospitable enclosures. The surfaces of these droplets permit communication and exchange between inside and outside. | Absorbing solar energy, as well as organic matter from Earth, comets, and asteriods, these pre-life forms become increasingly complex. Growing, maintaining, and self-regulating, they transform subtly, amazingly, into living cells.

25

*Billions of bytes of satellite information allow us to view life today as a planetary phenomenon.*

(left) Extremist microbes inside tube worms form the cornerstone of this 20th-century deep-sea hydrothermal vent ecosystem.

(below) Extremists bask in this sulfuric-acid-rich river, Rio Tinto, Huelva, Spain.

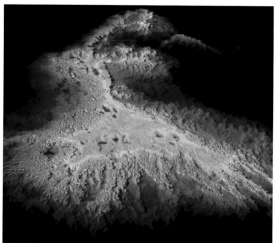

26

## 4000 MILLION YEARS AGO

### THE EXTREMISTS

Bacteria are Earth's first life forms. They continue to dominate environmental evolution in the twentieth-century. Ancient bacteria (archae) thrive in extreme environments typical of Archean (early) Earth. |

Heat- and acid-loving microbes flourish in boiling muds, hot springs, deep-sea vents, and ash-ejecting volcanoes. Some loll in waters as hot as 113 degrees C [235 degrees F]. Some freeze at temperatures as high as 55 degrees C [131 degrees F]! Some bacteria abide in pools of concentrated sulfuric acid. | Twentieth-century microbial methane makers thrive in oxygen-free sediments inside animals, and in sewage. Producing all the methane in Earth's atmosphere, these bacteria help prevent oxygen levels from reaching an explosive concentration.

plants and animals do not evolve until three-fourths of the way into the evolution story. Suddenly the old "tree of evolution" we know from our schoolbooks changes dramatically. Animals, fungi, and plants are no longer its main branches; rather, all three are merely the tips on a single branch of a tree composed of myriad microbes—creatures too small to see with the naked eye.[1]

Before our new wave of knowledge about our single-celled ancestors—bacteria and nucleated cells called protists—the bulk of evolution was as murky a "prehistory" as the three million years of human existence prior to what we call the Stone Age. Now, quite suddenly, we are unveiling this surprisingly cosmopolitan ancient (and modern) microworld. Tracing the urban lifestyles of ancient bacteria with all their technologies—from skyscrapers to compasses and electric motors, even to a World Wide Web of information exchange—is an amazing journey we will take in these pages. Popular science essayist and former head of Yale Medical School Lewis Thomas has even suggested that bacteria may have invented us as big taxis to get around in safely.[2]

## The DNA Revolution

The discovery of DNA structure in the 1950s has since engendered vast amounts of information about its role in individuals and in evolution, as well as the whole field of genetic engineering. Most notably, our view of DNA as a fixed "blueprint" in each creature, altered only by accidents in the course of evolution, is changing dramatically. We are still in the early stages of an exciting new view of DNA as a complex self-organizing system that responds to events outside the cells and creatures in whose development, maintenance, and evolution it plays such a central role.

## Creativity in Crisis

Even in ancient times, when bacteria faced food shortages, global atmospheric pollution, and destructive ultraviolet radiation, such challenges led to the invention of new DNA genes and new lifestyles. Today, bacteria respond similarly to each new generation of antibiotics, quickly making themselves resistant. Animals and plants later faced crises such as extinctions, the survivors of each such crisis retooling, evolving into new forms and functions. It is beginning to look as though crises afford life unusual evolutionary opportunities to create novel solutions.

These emerging themes—the geobiological systems view, our microbial ancestry, the DNA revolution, and the creativity of life in response to crisis—are leading us to a new story of evolution. In these pages we will look through a

27

macroscope at our story's larger patterns of evolutionary process even as we examine its microscopic detail.

Our Cosmic Prologue leaves off as the new planet Earth whirls into being—one of the Milky Way's supernova debris dustballs. Far out on one arm of our spiraling galaxy, in the third orbit away from the young star we call our Sun, Earth gathers itself together. It gets denser and its still radioactive elements form a heavy molten core while its lighter elements form a rocky skin around it, like a hot pudding as it cools.

## EARTH COMES TO LIFE: THE GREAT BIOGEOCHEMICAL CYCLES

What makes early Earth come to life while the other planets and moons of our solar system do not? What features does our planet have that enables it to evolve and support living creatures?

Several necessary conditions for Earth life are known. One requirement is a supply of needed elements. As heavier elements sink to the core, the new Earth gains mass and thus gravitational pull, enabling it to attract and retain light gaseous compounds of supernova debris in which the elements of life are found. Living creatures will primarily be composed of four elements: carbon, nitrogen, oxygen, and hydrogen, all of which now exist on Earth's surface.

Other necessary conditions for Earth life are a relatively narrow temperature range and sufficient sunlight. Earth warms itself from within and without, as radioactive elements from supernovae explosions generate tremendous heat and surface bombardment continues. But life also depends upon heat and light from the Sun reaching its surface. For the right amount of heat and light, Earth must be within a certain range of distance from the Sun. James Lovelock, the British atmospheric chemist who proposed that Earth is alive in his Gaia hypothesis, called this the "Goldilocks Effect": Earth had to be "not too hot, not too cold, but just right."[3, 4]

Life requires materials that move about, permitting cycles to develop. Living systems, for example, need continual access to new source materials and the means for moving used materials we call wastes to species that can use them for their own purposes. The movement of Earth's components also makes it possible to redistribute heat and the chemical composition of the atmosphere, waters, and minerals.

The mobile materials of Earth comprise three recycling systems that develop early in its history. In the first system, called the lithosphere —literally, "rock sphere"—magma erupts from below the crust to the surface. As it pours through cracks (now seafloor rifts) and volcanoes, it forms new cooled crust, as well as steam and gases. As the crust thickens from continual eruptions, it eventually breaks up into tectonic plates. We can identify their outlines today as earthquake and volcanic activity belts. While some plate edges spread with newly cooled magma, their opposite edges are pushed under adjoining plates and melted back into magma. Regions where this happens are called subduction zones.

Almost no rock on the surface of Earth today is original rock. Even granite almost four billion years old has already been recycled. Continents carried on the backs of tectonic plates move at an average rate of twenty miles in a million years. That means the continents altogether may have traveled more than 100,000 miles in the course of Earth's history, ever recycling and renewing themselves at their changing edges. Some scientists estimate that they have formed a single great landmass (the most recent called Pangaea) and broken up to form continents as many as four times in the course of Earth's history.[5]

Over evolutionary time, the lithosphere becomes inextricably interwoven with living species. Marble achieves its beautiful patterns from bacterial colonies; chalky cliffs are formed from tiny sea creatures, as is all limestone. Other microscopic creatures form continental shelves and reefs, while still more eat into rock, thus making land soils. It is extremely difficult to find any part of Earth's crust today that has not been altered and permeated by living creatures.

Ore deposits also show the important role of microorganisms, or microbes—interchangeable names for creatures too small to see with the naked eye—on the lithosphere over time. Many veins of ores exist because microbes have the chemical effect of precipitating minerals out of water, thus settling them to the bottom. Some microbes consume minerals, which then remain in their "resident" places as they die in huge numbers. According to Canadian environmental chemist William Fyfe, "For many elements . . . there is a good chance that they have spent part of their lifetime on the planet inside a living cell."[6] Colonies of microbes are found down to a depth of more than four kilometers inside Earth's crust. "As deep scientific drilling is developed, a host of observations show products from the deep biosphere. Indeed, if there is a cavity of appropriate size with sufficient water, life will be present. We must understand the deep biosphere if we are to correctly describe the carbon, nitrogen, and sulfur dynamics of Earth."[7]

The second recycling system is the hydrosphere—literally, the "water sphere." It begins with steam released from magma, which pours into the

29

atmosphere. The more Earth cools, the more the steam in its early atmosphere condenses into torrential rains. Streams form, join into rivers, and wash into pooling seas. As the seas form, the surface waters also evaporate into clouds under the Sun's heat, creating a cycle of weather driven by solar energy.

Later in Earth's evolution, living systems assume a very active role in weather cycles. Hydrogen sulfide gas, for example, is produced by plankton covering large ocean surfaces. As it rises into the atmosphere, molecules of this gas form nuclei that attract water molecules and form raindrops. Thus, the plankton seed rain clouds. Whales, as huge consumers of plankton, may play a role in keeping cloud cover in check. Humans not only kill whales, they create agricultural chemical runoffs that feed huge populations of new ocean plankton.

Huge rain forests in the equatorial regions of Earth play another role. As rainwater is evaporated from oceans, it is swept to shore by ocean winds and falls on the rain forests, which pump much of it high into the atmosphere where it recycles to polar regions and falls as snow that eventually melts back into oceans.

The third recycling system is the atmosphere. The early atmosphere may have been hot and loaded with methane, ammonia, nitrogen, carbon dioxide, and water, though scientists differ on its composition. Large organic molecules are formed from such compounds by the intense energy sources of the Sun, of erupting magma, and of lightning storms. Today's atmospheric composition is dramatically different due to the activity of living systems that created it and continually replenish it, but it continues to cycle its gases as weather systems drive it and as it is breathed in and out of living systems. Every lungful of air we breathe has very recently been recycled through

30

---

### 3900 MILLION YEARS AGO
#### LIFE TAKES TENACIOUS HOLD

The chronometer marking life's origin continues to march backward as we develop new tools for observation and analysis. Recent discoveries suggest that life on Earth may originate as early as

3,850 MYA, even as the planet sustains heavy bombardment from meteorites and other incoming bodies.

*Archean earthscape*

31

2,000 MYA                    1,000 MYA                    PRESENT

other organisms with which we share our planet.

Understanding how these great recycling systems of Earth—the lithosphere, hydrosphere, and atmosphere—are interwoven with living organisms—the biosphere—gives us a new picture of evolution. We can no longer see Earth as an assembly of parts, as biological systems separate from geological systems, or as living creatures against backdrops of nonliving habitats. Earth is a highly mobile geophysiological planet that continually changes itself in the process we call evolution.

## New Ways of Looking at Life

Scientists have not found it easy to define life. Almost all attempts to do so focus attention on individual organisms rather than on the broader living systems. One exception is the Russian geologist Vladimir Vernadsky's view of life as "a disperse of rock."[8,9] Vernadsky sees life as a geochemical process: Earth's crustal rock, either solid or in the form of sand and dust, transforms into highly active living matter and then back into rock.

In Vernadsky's view, life is a kind of planetary metabolic activity, metabolism being our name for all of the biochemical cycles and processes of life. As crustal components are "packaged" into cells and multicelled organisms, chemical changes speed up. Bacteria break up, consume, and then move crustal materials. They and larger creatures turn solar radiation and each other into bioenergy. At death, the creatures themselves are reduced back into soil and sediments, completing the cycle as they return to rock. In short, life in Vernadsky's view is rock rearranging itself with the help of solar energy and water.

To illustrate this process with a dramatic example, Vernadsky points out that a locust

---

**3 8 0 0** M I L L I O N   Y E A R S   A G O

CELEBRATING DIVERSITY *Fast and Loose*

Fast bacteria divide, cloning themselves every 20 minutes. In a million divisions, one bacterium may be a mutant. While most mutants die, successful ones quickly clone themselves across the environment. | Bacteria aren't just fast; they're also loose. Gene traders and swappers, they do not just create the next generation—they can become the next generation. "Horizontal" evolution yields brand-new kinds of beings. | What would happen if human beings could swap ideas as readily as these bacteria swap genes?

*In Baja California, billions of phototrophic bacteria clone them-selves and swap their genes in the warmth of a sunlit salt marsh puddle. With adequate food, water, and space (and no predators!), a single bacterium could generate $2^{144}$ individuals in two days (vastly more than the total number of human beings who have ever lived) and, in four days, $2^{266}$ individuals (greater than the number of atoms physicists estimate to exist in the universe).*

plague of a single day has been estimated to fill six thousand cubic kilometers of space and weigh 45 million tons. This means that more than 45 million tons of soil and water were converted into plant matter, much of it suddenly consumed by locusts that die even more suddenly and are transformed back into soil by bacteria as they decay.

Vernadsky's view of life fits well with the planetary cycles we have described and helps us take a more holistic view of evolution, though most biogeochemical activity goes on less dramatically. Certainly it is interesting to consider what happens to atoms and molecules in the course of evolution.

If we track a single elemental atom of Earth —say an atom of silicon—through time, we might see its journey beginning deep in core magma. Tracing its path, we see it erupt to the surface, find it in rock, then see it eaten by bacteria. It moves on into various living creatures

*(below) Chromatium swim toward light and their favorite food, hydrogen sulfide gas, and swim away at the slightest whiff of oxygen. Like most bacteria in the wild, they live in interdependent communities of large and varied populations.*

*(opposite) This lake community shows contemporary relatives of Earth's earliest photosynthesizers. Green Chlorobium, the first solar-powered sulfide munchers of the planet, reside here with pink and purple populations of other early non-oxygen photosynthesizers. About 3,700 MYA, the Archean landscape glistened bright green, red, purple, and orange as hydrogen-hungry microbes colonized wet volcanic terrain, pumice, and black sands.*

34

| 3700 | MILLION YEARS AGO |

LIFE'S FORTUNATE FERMENTERS

Life is easy for the biosphere's first beings, fermenting a pantry filled with free organic compounds formed in the atmosphere. These anaerobic (living in the absence of oxygen) freeloaders create life's first food crisis: rapidly reproducing, they consume food faster than the atmosphere renews it. | Certain descendants of these fortunate fermenters show notable problem-solving skills: they overcome the food shortage by learning to make their own food. These planetary "primary producers" use light or chemicals to generate energy, fabricating food directly from carbon dioxide. | Green and purple microbes evolve Earth's most important metabolic innovation: photosynthesis. These early prodigies practice specialized photosynthesis, which gives off sulfur rather than

35

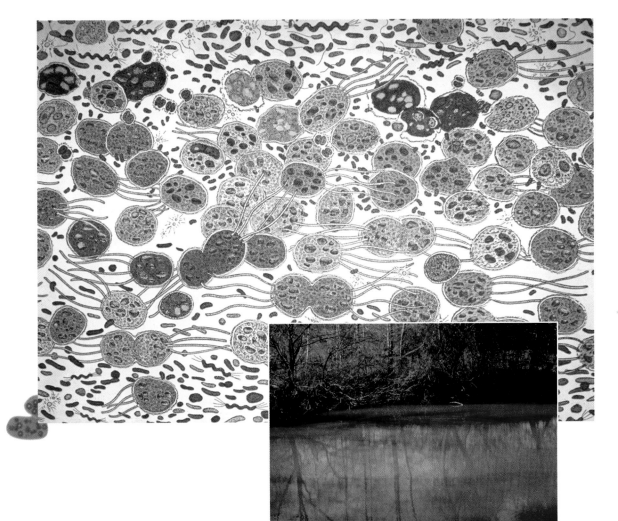

oxygen "waste." Using solar power, the microbes take hydrogen from hydrogen sulfide gas that is spewed out of deep-sea vents and volcanoes and combine it with carbon dioxide to make their bodies.

*Pollution – NOT! Virtually all lakes harbor descendants of ancient microbes which photosynthesize but which cannot tolerate oxygen and thus usually remain in murky depths. Here, at Lake Cisó in Spain, abundant hydrogen sulfide and trees protect the waters from aerating winds, and the healthy* Chromatium *turn the lake a comely pink.*

from microbes to trees and animals by turn, later moving about in water, evaporating into a cloud, falling on a mountain peak, and being carried by water back to the sea. There it recycles through more creatures before it ends in sediment and is possibly remelted into magma again. How shall we say whether it is animate or inanimate, geology or biology? Can it be that what we call life and nonlife are simply complementarities in the great dance of life over time, as physicists have found with mass and energy, particle and wave?

In 1937, ten years after the publication of Vernadsky's *The Biosphere*, British geochemist V. M. Goldschmidt wrote about the influence of the biosphere on geology. Meanwhile, G. E. Hutchinson at Yale University was promoting Vernadsky's view of life as a biogeochemical process of Earth. William Fyfe points out that the scale of this influence is only now being appreciated.[6]

While Vernadsky never stated that Earth itself is alive, that concept has been common among many cultures. Within the framework of western science, it was proposed at various times. Scottish scientist James Hutton, whom we honor as the father of geology, called Earth a living superorganism in 1785 and said its proper study should be physiology.

More recently, James Lovelock updated this idea in his Gaia hypothesis, which stated that living and nonliving systems of Earth are tightly coupled and that life creates its optimal conditions by regulating surface temperatures and chemical balance in soils, seas, and atmosphere.[3, 4] He named this living system Gaia, after the Greek goddess whose name means "Earth" in Greek, and whose ancient name, Ge, is found in our words geometry (the measure of Earth), geology (the organization of Earth), and geography (the mapping of Earth).

For a long time, as Lovelock pointed out,

---

## 3600 MILLION YEARS AGO

### TRIPPING THE LIGHT FANTASTIC *And the Waters Separated*

As Earth calms, hydrogen becomes scarce. The cyanobacteria (the blue-greens, wiliest of the microbes), pioneer the intranet approach: inside themselves, they link together two photosystems. This gives cyanobacteria enough energy to split tightly knit water molecules and procure the hydrogen for their bodies. | Waste Not. Great innovations are often characterized by unexpected novelties and unimaginable results. This power-plant technique of harvesting hydrogen also produces highly toxic waste, oxygen, which poisons other anaerobic microbes while opening new and diverse paths for oxygen-breathing biota.

Microcoleus cthonoplastes *bacteria, soaking in sunlight, employ protective sheaths to screen out ultraviolet radiation. We still do not understand how these gliding mat builders move.*

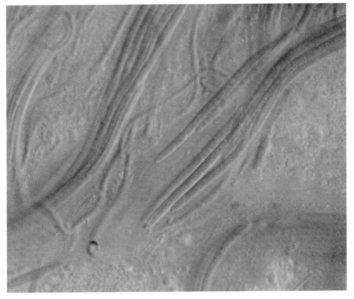

scientists really had no clear definition of life. Biology students simply memorized lists of the attributes of organisms: mobility, irritability, growth, reproduction, and so on. A real definition of life was at last proposed by Chilean biologists Humberto Maturana and Francisco Varela in the 1980s.[10] They called it *autopoiesis*—literally, self-composition or self-organization. The definition stated that an entity is alive if it creates and continually renews itself, including its defining boundary (read: cell wall, skin, membrane, bark).

Note that this core definition of life does not include growth or reproduction, which therefore may or may not be features of a living entity. Might it apply to Earth? Earth continually creates itself anew from the inside outward, including the tight atmospheric boundary we see hugging it closely in photos taken from space. Lewis Thomas has suggested that Earth is like a giant cell—an interesting idea given that cells are the basis of all living entities within Earth.[2] It would seem that the definition of life as autopoiesis fits Earth well, but this remains a controversial idea.

Without doubt, evolution occurs as a complex and changing web of interactions. As we gain in our understanding of them, we see how inseparable are the "biosphere" and "geospheres" (lithosphere, hydrosphere, and atmosphere). Whole new fields of geobiology or geophysiology are developing. Only through a systemic view of evolution can we understand the worldwide network of microbes that evolved into the larger life-forms, always inseparable from the habitats they create. Such systemics are now moving to the forefront of evolution research and theory, which is vastly increasing our understanding beyond the Darwinian and neo-Darwinian theories that taught us our common ancestry in nature and engendered all the research that brought us to where our vision could expand.[11, 12]

38

## 3500 MILLION YEARS AGO
### STROMATOLITES *Community Living*

The benefits of community life impress microbes early on. One microbe's waste is another's lunch. Eating, reproducing, and making waste are consistent features in the continual development of life.

Microbial mats form richly layered ecosystems, and under the right conditions, these become stromatolite bacterial skyscrapers. | The blue-greens live in the top layers, slipping in and out of UV-light-shielding sheaths to gather solar energy. Cyanobacteria produce prodigious amounts of food. "Consumer" bacteria, immune to oxygen, quickly join the cyanobacteria. Beneath them live mixed populations of consumers and producers, each possessing unique diets and tolerances for oxygen, light, and sulfides.

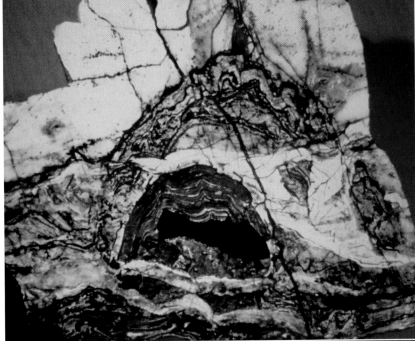

This fossil stromatolite (left) sings of the benefits of bustling, layered microbial community life in Warrawoona, northwest Australia, 3,500 MYA. The living microbial mat (below) is from Matanzas, Cuba.

## CREATIVE PROCESS IN
## TURBULENT TIMES

In the Archean era, Earth orbits a young Sun, less brilliant than now, as a hot, reddish-brown, largely molten, still radioactive planet cooling its crust like skin on a steamy pudding. As we move closer, we see its thickening skin heaving, cracking, and sliding about, pouring forth ever more of its fiery magma, pushing up mountains, and buckling in valleys, only to rearrange them again and again. Meteors continually strike the changing landscape, splatting down to form craters of all sizes while scattering about crustal materials.

Magma cools into rock while its gases and steam form a murky early atmosphere. As steam collects in vast quantities, the great rains begin, hissing to Earth and helping it cool. Lightning storms rage as electromagnetic energy gathers and is pulled to the surface. Eventually, rivers form and pour themselves into growing seas where they meet in lower landscapes. Erosion begins as cooled rock breaks up or is sloughed off by the rains, washing down rivers and into the seas, filling them with salts and other minerals.

Scientists are still divided on whether the planet cools to an icy surface pocked with volcanoes by the time living creatures emerge, whether it is temperate overall or still fiery hot.[1] Whatever the surface temperature, it is still a turbulent landscape. Nevertheless, creatures emerge. But how?

The key components of Earth's creatures are carbon and reduced carbon compounds, which are carbon atoms surrounded by hydrogen atoms. This is a lively, energized form of carbon that combines easily with oxygen, nitrogen, sulfur, and phosphorus to form organic molecules. We are all made of little other than these six elements in their rich variety of combinations.

We know that organic molecules exist in space, travel to Earth on meteorites, and are also forming on the surface of early Earth. In lightning storms, such molecules absorb electrical energy, which probably speeds up their chemical reactions to combine them into the larger molecules of sugars, acids, and lipids (fats). Various combinations of a dozen or fewer carbon, nitrogen, hydrogen, and oxygen atoms form amino acids, which in turn form long chains. This much has been shown possible in laboratory simulations of presumed Archean conditions, beginning with Stanley Miller's work in 1952[1], but exactly how the giant molecules that are needed to build living creatures formed is still subject to debate.

The world of organic molecules is exuberantly creative. They find endless ways of linking and rearranging themselves into patterns. Some are catalysts helping others join together; many

40

different molecules loop themselves into cycles of chemical reactions. All this is going on in the probably fiery Archean world long before full-fledged cellular creatures emerge.

Among the giant molecules formed from smaller ones are long strings of amino acids we call proteins. Proteins become the main building material of living creatures. Certain protein molecules come to play a particularly important role by speeding up the chemical reactions—or interactions—of other molecules. These special proteins are enzymes, and their wonderful talent for speeding up the chemical processes of living matter—its metabolism—is crucial. The presence of enzymes has even been suggested as one way of identifying the presence of life.

Other giant molecules, assembling from both acids and sugars, are those we call RNA (ribonucleic acid) and DNA (deoxyribonucleic acid). DNA may actually have been a later development of early living systems based on RNA. Both DNA and RNA come to play key roles in the information systems that direct the building and maintenance of organisms.

Some scientists believe that such giant molecules first form as amino acids or other large molecules attach themselves to the regularly repeating surface patterns of crystalline matter such as clay or ice. Others believe that giant mol-

ecules form only after the earliest molecular life systems are already organized within tiny capsules that protect them from being dissolved.

Such tiny capsules, visible only with electron microscopes, are called liposomes—meaning "fat bodies." They form as hollow spheres, like microscopic soap bubbles, whenever lipid molecules find themselves in water. Lipid molecules have tails that are hydrophobic, meaning they avoid water. Therefore, whenever lipid molecules find themselves in water, they swing their tails inward as their heads form a tight protective sphere around them. Sometimes a double-layered membrane of lipid molecules forms a liposome bubble, with all the tails inside the double layer of heads. Such liposomes can hold water inside as well as tolerating it outside. Whether liposomes played an important role in the formation of the first cells or not, this is the typical formation of simple cell membranes and persists as the basic structure of complex cell walls.[13, 14]

Biophysicist David Deamer has found that organic compounds from meteorites "plump up as capsules" when exposed to tide-pool conditions.[1] Some biologists believe that soups at the edges of Archean seas contained liposomes and a variety of large molecules, all of which repeatedly dried out and liquefied again. This could cause the liposomes to break open and flatten out during dry times. Then they would form their spheres again in wet times, trapping large molecules—even as large as DNA and protein molecules—inside them.

It is easy to imagine such conditions at the edges of Archean seas. Liposomes could thus form the first cell membranes. Cell membranes protect the molecules inside them and at the same time connect them to the outside world by selectively permitting some kinds of molecules to come in and others to pass outward through them. This soon makes the inside environment chemically different from that outside. Such a protected situation might foster the development of elaborate chemical cycles that are basic to living cells.

At some point—in liposomes or on their own—DNA and RNA work together with proteins as the information-storing, copying, and building systems of life. The formation of

*Bacteria practice many colony styles.*

### 3400 MILLION YEARS AGO
#### LIFE STYLES OF THE LITTLE

As the thickest and oldest parts of the continents form, microbes experiment with various lifestyles. Those who take up swimming explore habitats, moving with grace from meal to meal. Many microbes pursue colonial life styles, huge populations mixing with one another for food and flexible gene exchange. | Microbes also experiment with multicellular lifestyles, some forming complexes eerily like trees and other organisms of our familiar landscape.

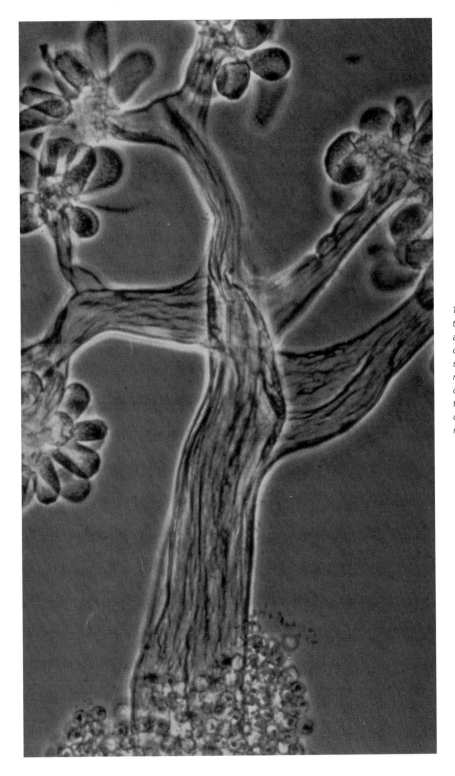

This myxobacterium practices the aggregate "tree" habit as a way to hang out and hang on through nutrient and water scarcities. When resources rebound, this fancy microbe drops "melon"-like pseudo fruits, releasing thousands of bacteria to procreate and recreate.

cooperative partnerships among them may become inevitable in shallow waters with the help of sunlight and lightning storms, or around cracks in the seafloor crust with the help of Earth's internal energy, or perhaps in both kinds of places at once. Scientists do not know just how it happens, but protein becomes necessary to initiate the process of DNA replication and also induces DNA to store codes, or "blueprints," for building specific proteins.

How does this partnership work? DNA "macromolecules" are long, double-helical chains of smaller molecules called nucleotides, each of which is made of a sugar, a phosphate, and a base. We picture the sugar-phosphate parts, which are all alike, as forming the sides of a twisted ladder in which each rung connecting the sides is formed from two linked bases, one from each side of the ladder. These bases come in four types that are coded into sequences along the ladder like letters into words and books. The bases thus permit DNA to hold complex and specific information in its formation.

The four bases are adenine (A) and thymine (T), which are always linked together, and cytosine (C) and guanine (G), which are also always linked together. All four appear on each side of the ladder, but because of the pair links, one side of the ladder determines the code on the other side. Rungs are thus AT, TA, CG, and GC pairs.

A sequence of three bases is called a codon. Four types arranged into threes is $4^3$ or sixty-four possible codons. Each codon except for four is a template for producing one amino acid, the building molecule for protein macromolecules, including enzymes. The others are one start codon that always indicates the initiation of a "word" sequence and three that tell where it ends —like the actual spaces between our words. Many "words" are codon sequences for particular

## 3300 MILLION YEARS AGO
### GOLDILOCKS AND THE THREE ATMOSPHERES

Atmospheric differences among the inner planets are influenced jointly by their proximity to the Sun and individual aspects of their evolution. Earth is our solar system's "Goldilocks": Venus is too hot; Mars is too cold; Earth is just right. One of the big questions is: Why are the inner planets, so similar in origin and relative composition of elements, so different? What accounts for the strange nature of Earth's atmosphere? | Life makes the difference. Carbon-dioxide-consuming and -producing beings of the biosphere, tightly coupled with geological processes, have influenced Earth's faces and flows for thousands of millions of years.

44

## Planetary Atmospheres

*Percent by weight of oxygen, nitrogen, and carbon dioxide*

*Mean annual surface temperature in °C*

Global temperature, gaseous atmospheric composition, and oceanic salinity and alkalinity fall under active biological modulation. Chemical, geological, and geophysiological forces influence our environment. The biosphere forms a living body, an aggregate of interactive Earth.

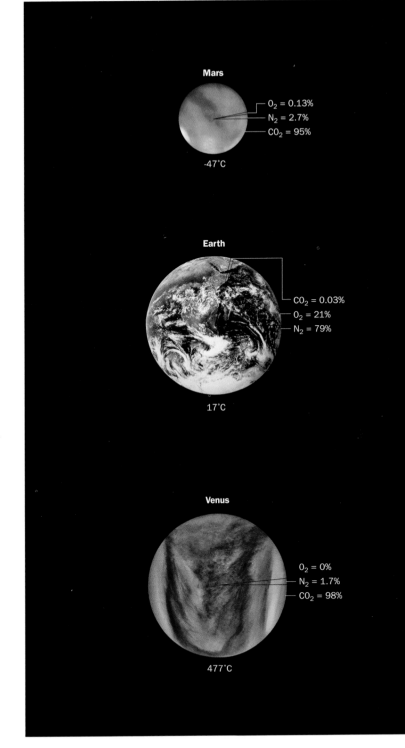

**Mars**

$O_2 = 0.13\%$
$N_2 = 2.7\%$
$CO_2 = 95\%$

-47°C

**Earth**

$CO_2 = 0.03\%$
$O_2 = 21\%$
$N_2 = 79\%$

17°C

**Venus**

$O_2 = 0\%$
$N_2 = 1.7\%$
$CO_2 = 98\%$

477°C

45

proteins, but the "language" of all DNA remains to be discovered, as most of it is still not understood.

Proteins are needed to uncouple the base links along one or more sections of the DNA ladder when, for example, the cell needs more of a particular protein. Once a section is separated, like a piece of open zipper, one half serves as a template for creating a piece of single-stranded RNA, which can detach itself to move about in the cell and serve in turn as a template for making the required protein. When the cell itself reproduces, the DNA ladder as a whole separates and each half becomes a template for making its other half exactly like the original. This way the whole DNA-protein partnership can reproduce itself.

The protein-building materials of living creatures come in as many types as their DNA codes instruct for—around 100,000 in us. The code for each protein is a single gene and may be thousands of codons in length. Certain "regulatory genes" play roles in when and how other genes express themselves. Genes are related in families whose members are not necessarily physically close in the DNA. It takes many genes working together to produce all the proteins required for a visible genetic "trait" such as the tail of a bacterium or the eye of a fly.

---

**3200** M I L L I O N   Y E A R S   A G O

WE'VE COME A LONG WAY

Not until the 1950s do paleontologists begin to find signs of life earlier than the large-sized fossils dated 600 MYA. Suspecting that microbial life evolved before animals and plants, thoughtful researchers commence a search. | In South Africa, explorers find the Fig Tree Formation. A rare series of sedimentary and volcanic rocks many tens of thousands of feet thick, the formation contains some units of a dense rock called chert, a smooth form of quartz. Although the chert appears barren, a microscopic search reveals signs of the kind of life that helped make stromatolites. Fossilized bacteria, including cyanobacteria, lie peacefully arranged in smooth layers within the rock. Some of the minute microfossils

## EARTH'S FIRST CREATURES: APPEARANCE OF THE ARCHAE

Around four billion (4,000 million) years ago, the first full-fledged cells appear on the Archean landscape. They are creatures because they lead lifestyles we can infer from their chemical composition and their descendants, consuming and eliminating materials, repairing and reproducing themselves as wholes. They are, in short, bacteria of a special type we call archae—"ancient ones"—even though their lineages survive to this day. Archae still maintain lifestyles differing from those of other bacteria that branched off from them billions of years ago.

The first archae must be hardy pioneers to survive the violent upheavals, extreme temperatures, and harsh chemistry of early Earth.

47

*A layered chert from the Fig Tree Formation*

**within this formation constitute precious evidence of early life.**

All of today's bacteria closest to the most ancient lineages are thermophiles—literally, "heat lovers." This is what makes some scientists suspect the origin of life around magma-oozing seafloor rifts. Regardless of their precise origins, we find their descendants today near such rifts, in boiling muds at surface eruptions and in pools of pure sulfuric acid, as well as in less extreme muds. In any event, the first archae had to evolve below the surface to avoid the intense and destructive ultraviolet radiation of direct sunlight.

All bacteria are single-cell creatures whose technical name is prokaryotes, meaning "before kernel, or nucleus." This distinguishes them from the only other type of cell to evolve later: the much larger eukaryotes, meaning "with nucleus." The DNA of bacteria floats loosely inside them as a simple chain of genes, often forming a closed loop. In eukaryotes it is packaged by complex folding and housed in the nucleus.

Bacteria are composed of genetic material, chemicals arranged into complex metabolic cycles, various molecular infrastructures, and salt water, all contained within their selectively permeable cell membranes. We are still in the early stages of understanding their intricate and inventive lifestyles, as well as their importance to us.

The pioneering archae try out various patterns of consumption and elimination as they invent a variety of enzyme-driven metabolic cycles. They depend on a steady supply of free food—organic compounds such as sugars and acids created in the atmosphere and in surface waters, which also absorb their ejected wastes. Over time they try out all sorts of available compounds to grow themselves after reproduction, to repair themselves when they suffer damage, and to invent new lifestyles. When their innovations work well, they keep records of them in their DNA.

48

## 3100 MILLION YEARS AGO
### EARTH MOVES ON

Several huge moving and shifting crustal plates carry Earth's small continents about. These plates, of which the continents are the raised portions, are part of Earth's lithosphere, which includes Earth's crust and the top rigid layer of its mantle. The lithosphere overlays a deeper "plastic" layer of inner Earth, where high temperature and pressure prevent rocks from solidifying. | Plate movement continually generates fresh lithosphere as molten rock erupts at midocean ridges. Oceanic plate margins are drawn into the mantle and melt at continental edges. As these processes cycle, continents move together and pull apart, oceans shrink and expand, plates collide and form mountain ranges, and island chains appear and vanish.

*When a plate moves across a "hot spot," it leaves a volcanic trail.*

This is truly amazing and noteworthy: that they not only invent new variations of their lifestyles, but that they code for them in their genetic material and thus pass the inventions on to family, friends, and future generations. Their two ways of passing them on are by reproduction and by sex, which in Archean times are independent activities.

Their reproductive process is called *mitosis* and is accomplished by copying their DNA and pinching their cell walls inward to divide themselves into two parts, each containing a copy of the DNA. Our own body cells, except for our sex cells, reproduce in the same way. In mitotic reproduction, the "offspring" contain DNA only from the single parent cell.

Archean sex, on the other hand, only occurs between reproductive divisions. Sex from a genetic perspective is simply the fusion of genetic material from more than one individual in a single creature. Bacteria can literally rub up

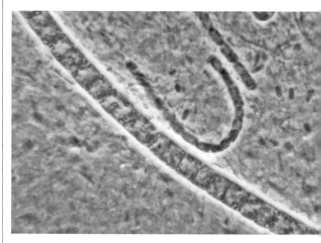

*This awesome blue-green bacterium is a great adapter, a switch hitter. Usually, it grabs hydrogen electrons from water, releasing oxygen. Under high hydrogen sulfide conditions, it stores gobs of yellow sulfur instead.*

50

## 3000 MILLION YEARS AGO
### METABOLIC MODES

Earth life forms require energy and carbon for growth and reproduction. They satisfy these requirements through a variety of processes. Some skilled individuals produce their own food; many rely on others to make it for them. | By this time, microbes have evolved every metabolic mode known to 20th-century scientists. Tiny gas-eating microbes, without using light, refine methods of making food and energy from sulfide, methane, ammonia, oxygen, and carbon dioxide. Cyanobacteria and their kin, using sunlight as a source of energy, create food from atmospheric carbon dioxide. Dependent on these primary producers, many microbes mix and match modes. | Stromatolite reefs, monuments of bacterial life, continue expanding across the planet.

This tall Canadian has stepped
back 1,800 million years, in
the Canadian Northwest
Territories, to show us just
how grand bacterial building
can be.

.

51

against each other, dissolve a common opening in their touching membranes, and slip DNA genes to each other. Bacteria have been doing this since they first evolved and continue the promiscuous practice to this day. Alternatively, they can release bits of DNA into the surrounding environment where other individuals can pick it up and assimilate it into their own DNA.

The fact that any type of bacterium can exchange genes with any other means that there are no species barriers among them. This mobility in the worldwide bacterial gene pool makes it impossible to identify bacteria clearly as species, so we refer to them as being of different "strains," a looser definition of type. Since the 1980s, some biologists have come to see them as a single superorganism.[15]

Bacteria can be killed or may die of environmental conditions, but they do not die of old age, for such death is a later development in evolution. Instead, they become their own succeeding generations by dividing as quickly as every twenty minutes or so. This can mean a billion bacteria from one parent in a day, and a billion more from each new one, and so on and on.

Because they are so highly prolific, archae soon spread far and wide. Their increasing appearance in Archean seas and ponds and on muddy shores becomes so massive a phenomenon

that it transforms the crustal surface itself. In Vernadsky's terms, the young planet's crust is transforming itself into layers of masses of microbial beings. In Maturana and Varela's terms, this is the autopoietic (self-organizing) activity of natural materials under the influence of natural energies. Some scientists call the bacterial masses covering Earth's surface the "cryptocrust" because it may not even be visible though it has enormous impact.

Just one example of this impact is the bacterial propensity for altering geology over time by rearranging and concentrating the minerals of Earth's crust, as described in the previous section. Consuming minerals in quantity, bacteria leave them behind as deposits. Alternatively, they precipitate, or settle, minerals suspended in water to the bottom. Today bacteria are actually used to perform such services in the mining industry.

Bacteria may have invented the atomic pile by concentrating radioactive materials, possibly to raise the temperature of their habitats by nuclear power. Today we find them as uranium deposits and use them for our own nuclear energy purposes. Billions of years of crustal rearrangement by bacteria has resulted in the formation of great veins of ores now mined by humans, as well as undersea carbonate platforms, reefs, atolls, and even continental shelves.

## CRISIS AND INNOVATION:
## THE ADVENTURES OF BUBBLERS,
## BLUEGREENS, AND BREATHERS

The diversity of metabolism (cell chemistry) in the bacterial world is far greater than that of all fungi, plants, and animals together. Much of what we know about bacterial evolution comes to us from the work of scientist Lynn Margulis and her research teams. [16, 17, 18]

Sorting out the different metabolic lifestyles of Earth's creatures in general has not been easy. The best way scientists have found to do so is by noting first that the energy and carbon needed to synthesize bacteria and all other life-forms comes from only two kinds of sources: direct (primary) or indirect (secondary).

The direct source of energy is light (*photo*), and its indirect source is chemical oxidation reactions (*chemo*). The direct source of carbon is $CO_2$ (carbon dioxide), and creatures that consume it are called *autotrophs*, meaning "self feeders." The indirect source of carbon is in more complex chemical compounds, and those who rely on them are called *heterotrophs*, meaning "other feeders." Summarized, we have four main categories of lifestyle, identified by what a life-form consumes in the way of energy and carbon:

- Photoautotrophs live off light and $CO_2$.
- Chemoautotrophs live off chemical oxidation reactions and $CO_2$.
- Photoheterotrophs live off light and complex compounds.
- Chemoheterotrophs live off chemical oxidation reactions and complex compounds.

While bacteria come in all four types, only two of these lifestyles get passed on to multicelled creatures later in evolution. Plants are photoautotrophs; fungi and animals are chemoheterotrophs.

An easier way to remember the most familiar lifestyles of ancient and modern bacteria is to use the mnemonic shorthand of calling them bubblers, bluegreens, and breathers.

Bubblers are fermenters that live by partially breaking down ready-made food molecules and emitting some of the remains as gases. They are named for the gas bubbles they make and can be found today in such places as swamps and yogurt.

Bluegreens (cyanobacteria) are named for their color and are photosynthesizers. They make their own food with light, carbon from $CO_2$, and hydrogen from water ($H_2O$). Their waste gas is oxygen. Bluegreens evolve into the chloroplasts in the cells of plants. Other photosynthesizers called sulfur bacteria consume hydrogen from hydrogen sulfide, produce sulfide globules, or leak

53

out sulfate ions. They come in many colors.

Breathers live by respiration, which is the use of oxygen, nitrates, or sulfates to oxidize carbon compounds to $CO_2$, which is their waste gas. Some of those that use oxygen evolve into the mitochondria in the cells of plants and animals.

We can see the symbiotic (literally, "together-living") loop set up between cyanobacterial bluegreens and oxygen-consuming breathers. The bluegreens' waste oxygen is needed by the breathers, and the breathers' waste $CO_2$ is food for the bluegreens. This symbiosis persists today in plants and animals. Perhaps native people intuitively recognize the value of their exhaled breaths to plants when they stand in the forest singing their gratitude for the plants' gifts to *them*.

Bubblers are first on the Archean scene, making their living as chemoheterotrophs, fermenting chemical compounds formed in their environments such as sugar and acid molecules.

54

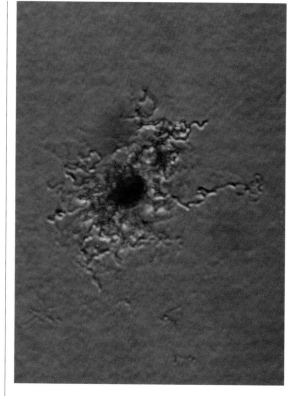

(above and right) Metal makers do their thing. These bacteria have precipitated manganese dioxide minerals out of solution.

<table>
<tr><td>2900</td><td>MILLION YEARS AGO</td></tr>
</table>

### MICROBIAL MINING AND MANUFACTURING

Bacteria modify the global mineral cycle in two important ways: they induce nearby minerals to precipitate (settle out of solution), and they internally manufacture minerals. Bacteria swimming in a river help precipitate the great gold deposits of South Africa. Today, these wee wonders head for the mines in Russia. Miners pump microbes and water into thin or hard-to-reach veins. | Initial research indicates that microbes biologically induce over 27 kinds of minerals. Which minerals these versatile fellows form depends upon their environment and their own genetic capabilities.

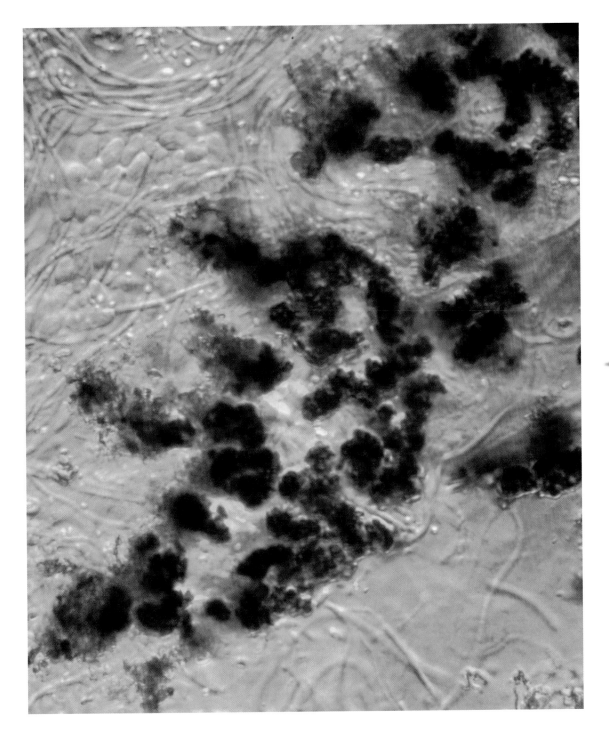

55

For example, they reduce sugars to alcohol, as we still get them to do for us today.

There is so much free food that they flourish, reproducing in vast numbers until they are consuming the available compounds faster than the environment can produce them. In short, these early creatures of Earth create its first global overpopulation and food crisis. In creating this crisis, however, they apparently drive themselves to new innovations, permitting life to continue. In fact, the food crisis leads to the evolution of all the other ways of making a living.

One innovation is to package DNA and protein partnerships into hard-shell spores that survive inactively for long times—floating or blowing about until better conditions permit them to grow back into full-fledged bacteria. Another is to continue trading DNA as information until new genes and new combinations of old genes produce new metabolic cycles and thus new lifestyles.

This permits bacteria to experiment with new foodstuffs. The elements they had found in free food, now depleted, are still all around them, but either hard to get at or not in usable combinations. Nitrogen, for example, is no longer readily available and must be "fixed" from other sources. Some bacteria learn to unlock the nitrogen from salty nitrates in the sea; others make atmospheric nitrogen usable by combining it with other elements. Had they not, life would have died out from nitrogen starvation, as nitrogen is one of the basic elements needed to build living things.

Another essential material now in short supply is carbon. This is the time some bacteria become chemoautotrophs by learning to unlock the carbon in atmospheric $CO_2$, thus making their own food from carbon dioxide and more complex chemical compounds.

Yet another innovation in response to the

56

## 2800 MILLION YEARS AGO

### HORDE HAVOC

Community living offers great benefits. One drawback is overcrowding, which leads to unneighborly behavior. Confusing friend with food, predators appear within the playfully swimming bacterial cooperatives. | The "Vampire-Berry" (*Vampirococcus*) attaches to and eats away the larger prey cells. Other microcosmic predators like *Bdellovibrio* slip just inside the surface of an unsuspecting neighbor, close the membrane and start digesting. A common predator lives and feeds by itself until resources diminish. When nearly overcome by crowding, it moves inside its intended victim and starts to divide. *Daptobacter* munches its red photosynthetic acquaintance and eats it all up. | It is speculated, in one of nature's luscious ironies, that the outcome of invasion, food crisis is the invention of photosynthesis, which means using light and other chemicals to make a living as photoheterotrophs. Bluegreen photoautotrophs come later. Pioneering photosynthesizers extract hydrogen from hydrogen sulfide, which spews from volcanoes and deep-sea magma eruptions that also produce the light they need.

Sulfur gas waste produced by the first photosynthesizers gives rise to another lifestyle: respiration. Sulfur is the first gas breathers use to break down organic compounds all the way to $CO_2$.

Hydrogen-hungry bacteria soon spread over the sunlight-rich surface in their many colors—green, pink, purple, orange, red, and yellow—giving a new look to Archean landscapes. Purple, green, and bluegreen bacteria find ways to harness light-sensitive chemicals such as the porphyrins that make our blood red and the chlorophyll that makes grass and leaves green.

**followed by truce, produces a grand blossoming in the history of life.**

Vampirococcus, Bdellovibrio, *and* Daptobacter *(top to bottom) practice new nutritional modes.*

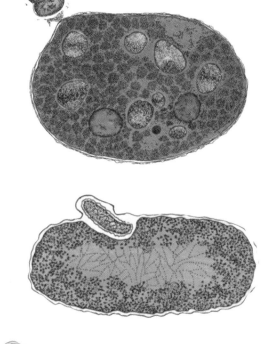

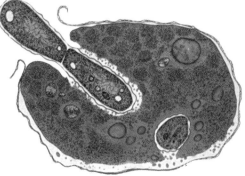

57

Photosynthesizers drive themselves to another innovation, perhaps as they deplete hydrogen sulfide reserves, already shrinking as Earth's crust grows thicker with cooled magma and eruptions slow down. Bluegreen cyanobacteria find a way to link two types of photosynthesis into one, gaining enough energy to split water molecules for their hydrogen. They are the first photoautotrophs—breaking down carbon dioxide, water, and rock salts into atoms, they reassemble them into food sugars. Bluegreens rapidly become the most successful bacteria of all.

Trapping sunlight and using it in making food is a truly marvelous step in evolution. Earth life today is still driven by solar energy because of this important invention. It is interesting to compare what happened to our bacterial ancestors with what is happening to us today. When pressed for energy to build human societies, we began to make much of it from coal and oil supplies found ready-made in our environment. Now these supplies are running out, and we must find new ways to produce energy. Among these new ways is solar energy, precisely what our photosynthesizing forebears learned to use.

Learning to extract hydrogen from the plentiful water all around them surely contributes to the huge success of the bluegreens. But all is not rosy in the Archean world when bluegreens take over. In extracting their carbon from $CO_2$ and their hydrogen from $H_2O$, they leave oxygen behind as their waste gas, and oxygen is very deadly.

Oxygen causes corrosion by breaking up and hooking onto many molecules it comes in contact with—a process called *oxidation*. As the highly prolific bluegreens spread over Earth, they produce it in such quantity that it rusts crustal iron ores, giving them a reddish color all over the planet. It also kills off many oxygen-intolerant

---

C O N C E P T       N O T E

SYMBIOGENESIS *Presence of the Past*

**Symbiosis** is the prolonged physical association of two or more different kinds of organisms. **Endosymbiosis** involves one life-form actually living inside another, often in a long-lasting merger.

**Symbiogenesis**, the source of true evolutionary novelty, occurs when the mergers of independent organisms actually form composites. A totally new kind of being may then evolve. | All life-forms (with the exception of bacteria which make them) are consortia. Our cells are the result of millions of years of mergers, the outcome of once free-living bacteria that came together in permanent relationships. | We are all chimera, composites of many life-forms and many mergers. Identity is less an object than a process. We all envelop traces of bygone beings.

*Images of Earth from the Moon vibrantly convey the life of the planet.*

59

"*The trading by bacteria of genetic information provides the basis for understanding new concepts of evolution. Evolution is no linear family tree, but change in the single multidimensional being that has grown now to cover the entire surface of Earth.*"

Lynn Margulis and Dorion Sagan

(anaerobic) bacteria. Some anaerobes burrow deeper into oxygen-free (anaerobic) environments such as muds.

When the oxygen has combined with, or oxidized, all the crustal materials possible, it begins to pile up in the atmosphere. Even the bluegreens have to invent enzymes that make the oxygen they produce harmless to themselves. Bluegreens also have to protect themselves from the ultraviolet (UV) radiation in the very sunlight on which their livelihood depends. For this, they cleverly invent UV-light-shielding sheaths they can slip in and out of as hermit crabs do with their shells. Some of these long, flexible sheaths are built cooperatively to house whole communities. What is perhaps even more interesting is the fact that bluegreens evolve the ability to repair genes damaged by UV light. It is only recently that microbiologists discovered our own genetic repair mechanisms.

60

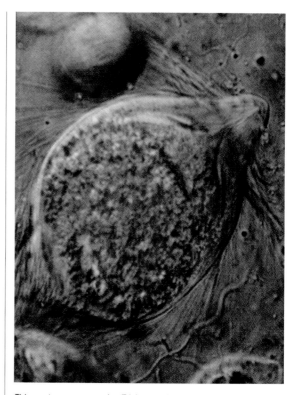

This contemporary protist, Trichonympha, *is pushed through its viscous termite hindgut habitat by thousands of symbiotic spirochete bacteria attaching at the rear.*

## 2700 MILLION YEARS AGO
### JOINT VENTURES

Special joint ventures occur in communities of mixed populations. A sluggish, ancient fermenting bacterium and a small, swimming, spirochete-like bacterium may have formed a particularly brilliant partnership. | Spirochetes, speedsters of the microbial world, arrive quickly at food sources. Their corkscrew bodies move perfectly through seaside muds, the viscous insides of animals, and all around our gums. Spirochetes have neither head nor tail until they attach to something. Seeping a sticky substance, individuals and often groups easily tack directly onto a larger microbe. | The adhered spirochetes enjoy the microbe's byproducts in exchange for providing their partner with fast and easy transport toward food.

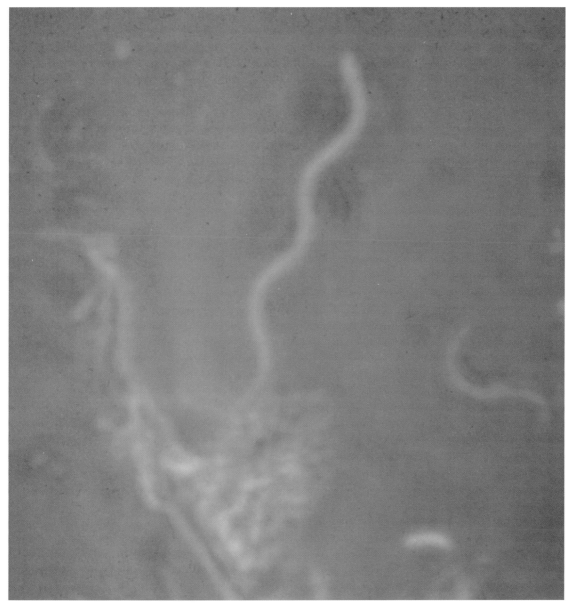

*Spirochetes are masters of movement, corkscrewing about with neither head nor tail.*

Our bacterial ancestors eventually invent every metabolic lifestyle known today. They also invent a single "energy currency" common to all life-forms, called *adenosine triphosphate*, or ATP. Breaking up molecules, whether by fermentation, respiration, or in the course of photosynthesis, frees the energy that held them together. Archae learned to store this energy in ATP molecules, which may at first have been found ready-made in their surroundings, though eventually they learn how to make them.

All the ancient bacteria store the energy-loaded ATP until its energy is needed for digestion, building, repair, and other work. Bluegreens are more efficient than the earlier-evolving bubblers at making their ATP and, in turn, breathers are even more efficient. Having this common energy currency surely facilitates the evolutionary mergers they are soon to make.

Those who wrest the greatest opportunity

### 2600 MILLION YEARS AGO

#### SURFACING

Larger, thicker, stronger continents emerge as Earth cools, closing the period of major crust formation. The continents grow and stabilize. Shallow, wide continental shelves provide ideal habitats for the growth and preservation of stromatolites. These cosmopolitan communities grow abundantly and luxuriantly along the continental margins.

62

*Miniature versions of deep-time–scapes are preserved in Shark Bay, Australia, and off the coast of Lee Stocking Island in the Bahamas.*

from the global crisis of oxygen pollution are the breathers, which harness oxygen's power to break up their food molecules. Being most efficient in making their ATP energy currency this way, they spend it on new transport technologies. Breathers earlier devised the spiral form of *spirochetes* for themselves, rotating their way through waters and muds as rapid corkscrews. But now a stupendous new invention arrives on the scene.

Breathers evolve a tiny spinning wheel at their tail end that rotates at fantastic speed in an electrical flow of hydrogen atoms called a proton motive force. They then attach one or more long, wavy flagella, or tails, to the rim of the wheel, enabling them to charge through water faster than any other moving microbe. [18] Thus, they invent not only the wheel but the electric motor in the microworld of three billion years ago.

## Urban Lifestyles and the First World Wide Web

All the long time that bubblers, bluegreens, and breathers are inventing diverse lifestyles and rearranging the planet's crust, they continue to exchange DNA information with each other. We can rightly say that they invent the first World Wide Web of information exchange. The importance of this astoundingly flexible gene pool cannot be underestimated. It is still as active among bacteria today as in Archean times and is related, for example, to their rapid resistance to our antibiotics.

Information exchange gives bacteria close relationships that facilitate cooperation in communal living. We have known of their communal lives for some time, but only now are we able to investigate their amazing urban complexes in real detail.

64

| 2500 MILLION YEARS AGO |
| --- |
| BIOLOGIC AND GEOLOGIC SYSTEMS WED |

**Modern geological processes begin. Coupling tightens between geologic and biologic systems. Mineral-forming organisms modify the chemical and physical nature of the biosphere. Bacteria induce** **minerals to precipitate huge carbonate platforms, giant barrier reefs, fringing reefs, and mounds.**

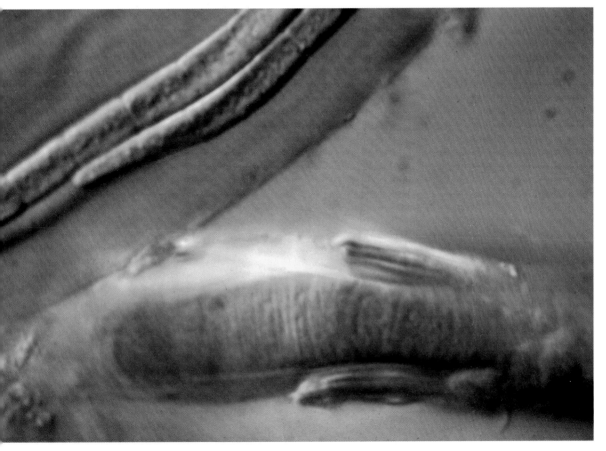

Mat, stromatolite, and reef-
building communities are highly
varied. These tiny architects
build huge and diverse rock-
solid parts of the planet.

In what seems to us the almost unthinkably ancient past, when bacteria still have the world to themselves, they not only discover the advantages of communal living but even evolve sophisticated cityscapes. We can see their huge urban complexes today. But with the naked eye they appear only as slimy films, thick muddy microbial mats, and giant fossilized communities called stromatolites —rocky domes of layered ancient seashore communities that trapped sand and other particles. Living slime cities persist on their surfaces.

Stromatolites are found in many locations, some pushed under the surface into fossilized banded rock formations, again reminding us of Vernadsky's definition of life as a transform of rock that goes back again to rock. Some stromatolites are still growing themselves on the surface in shallow waters and on shores. Other communal life experiments have less rigid forms than stromatolites. Some bacteria even create communities that look and sometimes act remarkably like later multicelled plants. Yet others adopt free-swimming lifestyles. One way or another, they all maintain community through their exchanges of resources and information.

Bacteria living on top of microbial mats or stromatolites are burned to death by ultraviolet light, but the dead cells make good filters, absorbing the burning rays while letting the rest of the light reach those that need it below. In other community situations, some individuals commit suicide so that others may live—a process called *apoptosis*, also found later in evolution as the embryological process of "programmed death," in which certain cells must die for multicelled creatures to "sculpt" their forms.

Bacterial cityscapes exist today wherever they can take hold—in wetlands, in dank closets, in the stomachs of cows, in kitchen drains. Scientists call them biofilms or mucilages, as they look like slimy brown or greenish patches to the unaided human eye. Only now can we discover their inner structure and functions with the newest microscopy techniques, which magnify them sufficiently without destroying them (for example, confocal scanning laser microscopy).

Looking closely for the first time at intact bacterial microcities, scientists are amazed to see them packed as tightly as our own urban centers, but with a decidedly futuristic look. Towers of spheres and cone- or mushroom-shaped skyscrapers soar one hundred to two hundred micrometers upward from a base of dense sticky sugars, other big molecules, and water, all collectively produced by the bacterial inhabitants. In these cities, different strains of bacteria with different enzymes help each other exploit food supplies that no strain can break down alone.[19]

All of them together build the city's infrastructure. The cities are laced with intricate channels connecting the buildings to circulate water, nutrients, enzymes, oxygen, and recyclable wastes. Their diverse inhabitants live in different microneighborhoods and glide, motor, or swim along roadways and canals. The more food available, the denser the populations become. Researcher Bill Keevil in England, making videos of these cityscapes, says of one, "It looks like Manhattan when you fly over it."[20, 21]

Microbiologist Bill Costerton in Montana observes: "All of a sudden, instead of individual organisms, you have communication, cell cooperation, cell specialization, and a basic circulatory system, as in plants or animals . . . It's a big intellectual break."[21] Researchers now are coming to see colonial bacteria or even all bacteria as multicelled creatures.[22]

Some bacterial colonies cause infections and diseases and deteriorate our teeth, our buildings, and our bridges. But most bacterial cooperatives are harmless or indeed cooperative with other creatures, many living inside their guts, as in termites and cows, helping with digestion. They maintain our worldwide habitats by renewing and chemically balancing the atmosphere, seas, and soils; they work for our health by the billions in our guts and have evolved into the organelles inside our cells.

We use bacteria (and yeasts) in our original biotechnologies for making cheese, yogurt, beer, wine, bread, soy sauce, and other foods. We harness bacteria for newer biotechnologies: to remove contaminants from water in sewage treatment plants, to clean up oil spills and other pollution, to refine oil, to mine ores, and even to make the biodegradable polyester they were making long before we were. All of our genetic engineering efforts depend on bacteria, as they do much of the work of DNA recombination in our laboratories.

Most astonishing to investigators, communal bacteria turn on a different set of genes than their genetically identical relatives roaming independently outside of biofilms. This gives the urban dwellers a very different biochemical makeup. A special bacterial chemical, *homoserine lactone*, signals incoming bacteria to turn into city dwellers. All bacteria constantly discharge low levels of this chemical. Large concentrations of it in urban environments trigger the urbanizing genetic changes, no matter what strain the bacteria are.

These changes include those that make bacteria most resistant to antibiotics. Costerton estimates that more than 99 percent of all bacteria live in biofilm communities, and finds that such communities, pooling their resources, can be up to fifteen hundred times more resistant to antibiotics than a single colony.[20] Under today's siege by antibiotics, bacteria respond with ever new genetic immunity. Our fifth generation of antibiotics failed in 1996.

In Tel Aviv, Eshel Ben-Jacob also finds bacteria-trading genes and discovers complex interactions between individuals and their communities. The genomes of individuals—defined as their full set of structural and regulatory genes—can and do alter their patterns in the interests of the bacterial community as a whole. He observes that bacteria signal each other chemically, calculate their own numbers in relation to food supplies, make decisions on how to behave accordingly to maximize community well-being, and collectively change their environments to their communal benefit.[23, 24]

Bacterial communities thus create complex genetic and behavioral patterns specific to different environmental conditions. The genomes of individual bacteria alter their composition, arrangement, and the pattern of which genes are turned on in response to changes in the environment or communal circumstances. This important information is coming from various research laboratories. Both Ben-Jacob and Costerton see individual bacteria gaining the benefits of group living by putting group interests ahead of their own. Ben-Jacob concludes that colonies form a kind of supermind genomic web of intelligent individual genomes. Such webs are capable of creative responses to the environment that bring about "cooperative self-improvement or cooperative evolution."[24]

Einstein's worldview was challenged when some quantum physicists suggested that electrons intentionally leap orbits.[25] Microbiologists are beginning to see similar intentional activity at

| 2 4 0 0 | M I L L I O N   Y E A R S   A G O |
|---|---|

ORE-IGINALS

Although global cyanobacteria release massive amounts of oxygen "waste," no surplus builds up in Earth's atmosphere. Oxygen gas reacts immediately with hydrogen, carbon, and iron to form oxides such as water, calcium carbonate (limestone), and iron ore (hematite and magnetite). | For 600 million years, sediments of alternately higher and lower concentrations of iron oxide settle out on the ocean floors. Some of these sediments metamorphose into massive banded-iron formations, or BIFs, which are the principal source of iron mined by humans two thousand million years in the future. | Scientists theorize that fluctuations in these iron oxide deposits were caused by some combination of seasonal upwelling of iron-rich waters from the depths of the ocean, seasonal

The "World-Wide Age of BIFs" extends from 2,400 MYA to 1,800 MYA. BIFs cover vast stretches of the planet, hundreds of kilometers across, and contain more than 90% of Earth's minable iron. This is a Jaspilite-hematite-quartzite BIF from Michigan, USA. The original iron sediments have been heated and folded under pressure. Next time you drive your car, you might want to thank the kindly Earth and its greatest bacteria, the bluegreens, for their contribution.

69

variations in photosynthetic activity, periodic volcanic eruptions, and seasonal variations in the oxygen production of cyanobacteria. | One interpretation is that the layers may be read as growth rings, indicating increases in oxygen "exhaust" by cyanobacteria in warmer seasons and lower exhaust in cooler seasons.

systemic, cellular, and molecular DNA levels. These discoveries of genomic changes in response to an organism's environment, in the context of a systems view of evolution, are changing our story of how evolution proceeds in significant ways. We are discovering, in short, that the fundamental life-forms from which all other organisms evolve are capable of both self-organization in community and self-improvement through environmental challenge.

Genomic changes in response to an organism's environment have actually been known since the 1950s, but they challenged the accepted theories of the time, so it has taken half a century to amass sufficient data to warrant changing our scientific picture of evolution accordingly.

Barbara McClintock, who did much of her work on corn plants, pioneered the research showing that DNA sequences move about to new locations and that this genetic activity increases

70

## 2300 MILLION YEARS AGO
### ENERGIZED

Able to swap and repair genes, some microbes evolve ways to tolerate oxygen by forming protective enzymes. These enzymes react with dangerous radicals produced by oxygen, converting them to innocuous compounds. | Other microbes develop a radical approach to oxygen, which both protects them and provides a powerful new means of energy transformation. They consume the oxygen produced by photosynthesis. Aerobic respiration commences: controlled combustion breaks down organic molecules for energy and gives off energy-poor carbon dioxide and water. | This innovation energizes life. Fermenting a single sugar molecule produces two molecules of ATP, the primary energy carrier for cell metabolism and motility. Processing the same sugar molecule through respiration yields

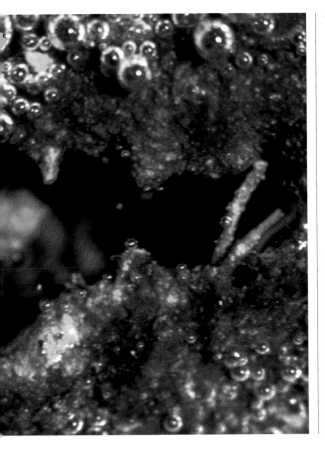

when the plants are stressed. She also found closed-loop molecular bits of self-reproducing DNA called plasmids moving about among the normal DNA and exchanged from cell to cell.[26, 27] Plasmids, which were invented by ancient bacteria and persist in multicelled creatures, are used a great deal in genetic engineering, as they can be inserted into new genomes.

McClintock's work on transposable genetic elements was verified and elaborated by many researchers until it became clear that DNA reorganizes itself and trades genes with other cells, even with other creatures.[13] The trading process sometimes involves viruslike elements known as transposons. Some are retrotransposons and retroviruses that transcribe their RNA into DNA—opposite to the usual order and not thought possible before their discovery. Some theorists now believe that bacteria may have invented viruses as well as plasmids.

71

**as many as 36 energizing ATP molecules.**

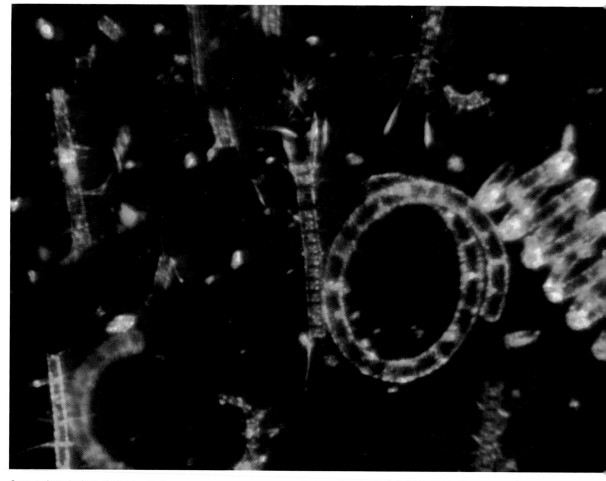

*Current photoplankton photo opportunity*

## 2200 MILLION YEARS AGO

### OFF TO SEA

Expanding reefs are a springboard for life in the oceans. Some bacterioplankton and algae, floating in the upper layers of Earth's oceans, become primary producers (photosynthesizers that make their own food) for expanding life. As they did in the shallows of early Earth, these microbes evolve a myriad of independent and interacting life styles.

Biologists Phillip Sharp and Richard Roberts, 1993 Nobel Laureates, discovered that RNA is arranged in modules that can be reshuffled by "spliceosomes," referred to as a cell's "editors."[28] Other researchers have shown that bacteria genetically retool themselves and can correct defects created by human genetic engineers.[29] (Recall that ancient bacteria had already evolved the ability to repair genes damaged by UV radiation.)

Further research shows that bacteria not only alter genomes very specifically in response to environmental pressures, but also transfer the mutations to other bacteria.[30, 31] Many of these genetic transfers appear to be evolutionarily related to "free-living" viruses, according to Temin and Engels in England.[32] Retroviruses are known to infect across species and enter the host's germline DNA.

We are still in the early stages of understanding the extent to which DNA is freely traded in the world of microbes to benefit both individuals and their communities. And we are just beginning to see these processes of genetic alteration at cellular levels as intelligent responses to changing environmental conditions in multicelled creatures.[13, 26, 27, 28, 31, 32] We know viruses and plasmids carry bits of DNA from whales to seagulls, from monkeys to cats, and so on, but it remains to be understood how much of this transfer is random and how much is meaningful.

Most research in this area is still confined to microbes in which these matters are easier to study. As yet we do not know to what extent DNA trading occurs in creatures larger than microbes, nor to what extent it facilitates specific responses to environmental conditions. For that matter, we still do not know what the vast proportion of the multicellular creature DNA does at all. Depending on the particular plant or animal species, only one percent to five or ten percent (in humans) codes for proteins. The remaining 90 to 99 percent remains a mystery! Even the much discussed human genome project is only concerned with mapping the protein-coding portion. So our stories are far from complete, but it seems reasonable to hazard the guess that nature would not have evolved an evolutionary strategy as sophisticated as gene trading to facilitate evolution billions of years ago only to abandon it in evolving larger creatures.

British researcher Jeffrey Pollard reports the rapid restructuring of genomes in response to stress in many different species, from microbes to plants and animals, with the changes passed on to succeeding generations. This can bring about, as Pollard says, "dramatic alterations of developmental plans independent of natural selection," which

73

itself may "play a minor role in evolutionary change, perhaps honing up the fit between the organism and its environment."[33]

This growing body of evidence suggests that evolution may proceed much faster under stress than was thought possible. It also reveals how the World Wide Web of DNA information exchange invented by archean bacteria still functions today, not only among bacteria, as always, but also within multicelled creatures and among species. As Lynn Margulis puts it: "Evolution is no linear family tree, but change in the single multidimensional being that has grown to cover the entire surface of Earth."

## OXYGEN CRISIS, OXYGEN SOLUTIONS

As oxygen created by bluegreens continues to pile up in the atmosphere, it not only causes problems that spur bacterial evolution into new lifestyles, it creates a significant benefit to life as a whole. The addition of oxygen to the other gases of the atmosphere creates a thick blanket of air. This slows the speed of incoming meteors with friction, which also heats many of them enough to burn them up before they can strike the ground. The more oxygen there is, the more meteors burn up, until so few of them get through the atmosphere that life is much safer on Earth.

A second benefit of the oxygen-rich atmosphere is the creation of the ozone layer. Atmospheric oxygen is composed of twin oxygen atoms, so chemically it is $O_2$. As ultraviolet rays strike the twin oxygen molecules, they break up the pair, leaving separated twins to join other oxygen pairs as triplet molecules. Such triplet molecules are no longer oxygen gas, for $O_3$ is ozone. Ultraviolet rays have difficulty passing through ozone because it absorbs them, so ozone

---

### 2100 MILLION YEARS AGO
#### PUTTING THE O'S INTO OZONE

As cyanobacteria-generated oxygen continues to accumulate in the atmosphere, a protective ozone layer begins to form. In another remarkable twist in the evolutionary story of life, oxygen, once a fatal form of air pollution, becomes a shield for future life against the sun's destructive ultraviolet (UV) rays. | Long before Earth's ozone layer formed, bacteria had developed mechanisms to cope with UV light: making sheaths, submerging themselves, digging in. They evolved the ability to repair genes damaged by UV light. Had no UV-absorbing ozone layer formed, life on Earth might have remained the exclusive realm of the microbes.

---

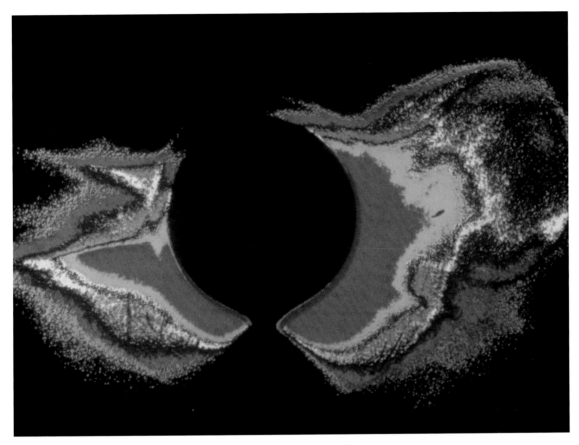

The solar corona, a halo of light around the darkened Sun, visible during a total eclipse.

*Year after year, layers of photosynthesizers and their followers pile up. This hand-sample of a contemporary mat community reflects over 100 years of stable community growth.*

## 2000 MILLION YEARS AGO

### ABUNDANT LIFE

Fossil records reflect the abundance of life. Total anaerobes bury themselves as deeply as they can in their communities. Strains of cyanobacteria, poisoned by their own oxygen waste, team up with populations of respiring bacteria which immediately slurp up the oxygen so toxic to their cousins. | Off and running in waves of innovation, aerobes' new-found energies allow them to experiment. They forge new niches, specialize, and try new sorts of lifestyles. Some bud motile offspring who, although clones, look nothing like the parent. These baby bacteria swim to a favorable location, attach to a solid surface, and "metamorphose" back into parental form.

is a natural sunscreen for the whole planet. As oxygen accumulates in the ancient atmosphere, a whole layer of ozone collects around fifty kilometers (thirty miles) above the surface, shielding Earth life from dangerous amounts of ultraviolet radiation. Without the ozone layer, it is unlikely that life would have evolved beyond bacteria.

James Lovelock's work on understanding the relationship between Earth's atmosphere and its living creatures gave strong support to his Gaia hypothesis that life creates its own life-supporting conditions.[3] Respiring creatures quickly recycle nearly all oxygen produced by photosynthesizing creatures, yet an exact proportion of oxygen—21 percent—is present in the atmosphere at all times. A little more and all vegetation, including wet grass, would catch fire so easily that every lightning storm would ensure raging conflagrations; a little less and the animal life dependent on it would cease.

The amount of carbon mined from atmospheric $CO_2$ by organisms and then buried in sediments after they die leaves a net increase in oxygen that is part of life's complex atmospheric balancing. Other critical roles in maintaining the oxygen balance precisely are played by gases such

77

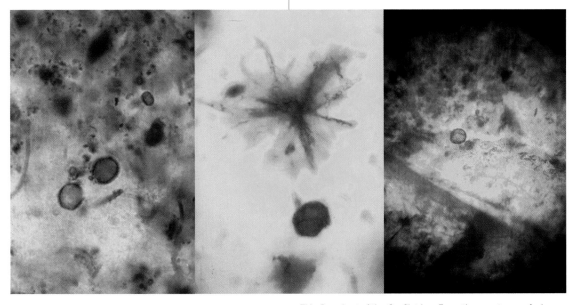

This fine chert of the Gunflint Iron Formation sports a profusion of stellate (star-shaped) and ovoid microorganism fossils. A complicated assemblage of bacterial communities have lain preserved, locked in the rocks of the formation for 2,000 million years.

as methane and ammonia produced by organisms and creating complex reactions with other atmospheric gases. Microbe-produced oxygen may even play a role in preventing the seas from drying up. Atmospheric oxygen traps lightweight hydrogen into water molecules, thus preventing the lightweight hydrogen from escaping into space and allowing it instead to fall back into the seas in the form of rain.

## THE MULTICREATURED CELL: A GIANT STEP IN EVOLUTION

78

The most remarkable cooperative venture since the evolution of urban life and the World Wide Web among bacteria comes together around two billion years ago, more than halfway through Earth's life, yet still in the Archean era. It is the evolution of the only kind of cell other than prokaryote (bacterial) ever to evolve—the eukaryote, or nucleated cell.

All creatures that are not prokaryotes are eukaryotes, or multicelled creatures composed of eukaryotes—fungi, animals, and plants. Single-cell eukaryotes are protists, meaning "first builders." Note that there are no single-cell plants or animals.

Remarkably, this newcomer on the Archean scene is actually a multicreatured cell, because it evolves from a consortium of different types of

bacteria within one cell wall. Lynn Margulis, who helped track this story, calls eukaryote cell evolution "symbiogenesis." The participants in the formation of the nucleated cell were bacterial symbionts.[16]

In 1910, a Russian scientist, K. S. Mereschovsky, proposed that "organelles" inside nucleated cells—particularly the mitochondria of plant and animal cells and the chloroplasts in plant cells—are descended from ancient free-living bacteria. Mitochondria occur abundantly in all our own cells, making our energy by breaking up food molecules with the oxygen we breathe as our bloodstream carries it to them. Chloroplasts, which belong to a broader category of plastids—photosynthesizing organelles—live cooperatively with mitochondria in plant cells and make food by photosynthesis.

Merechovsky's idea was brought to American attention in 1927 by Ivan Wallin in his book *Symbionticism and the Origin of Species*, but was virtually ignored until Margulis took it up in the 1970s.[34] Part of the evidence for cell symbiosis was the discovery that mitochondria and plastids, living inside eukaryotes but outside the nucleus, have their own DNA, separate from nuclear DNA. Also, they reproduce themselves, whether or not the cell in which they live is reproducing. All our mitochondria are descended from those of the egg cell with which we began, as they do not occur in sperm. Mitochondrial DNA is therefore referred to as maternal DNA.

Margulis and her research teams painstakingly puzzled out the long and complex steps through which ancient bacteria evolve into such integrated cell parts. The story that emerged is one of forming ancient cooperatives.[16, 17, 18, 34]

It seems to begin when corkscrew spirochetes, motor-driven breathers, set out aggressively to launch attacks on bigger, slower bubblers in the desperate search for new food supplies. Drilling their way into the sluggish bubblers, the breathers consume the bubblers as food, and sometimes stay to reproduce inside them until all supplies are exhausted. Bacteria such as *Bdellovibrio* still do this today. But destroying your host is a dead-end kind of colonialism, and some breathers apparently learn to stay inside bubblers without destroying them.

Apparently they discover the advantages of turning such exploitative colonialism into cooperative merger. Breathers are, after all, far more efficient at making ATP energy currency than are bubblers. The additional productive energy they can make may contribute to building stronger cell walls, thus permitting the breathers safer lodgings. Breathers also seem to help their hosts tolerate oxygenated environments in turn for shelter and sharing the bubbler's ingested food. In any case, breathers become cooperative members in these new communal living arrangements and eventually evolve into mitochondria. As Lewis Thomas has said, pointing out that they constitute half our dry weight (body weight minus water), they may have created us as giant taxis to get around in safely! From our own perspective, we could not lift a finger without them.[2]

Somewhere along the way, light-tolerant bubblers engulf smaller green or blue-green bacteria in their own hunt for scarce food. But rather than digesting them, some bubblers let them go on making food from the ever-available carbon dioxide and sunlight—enough food for themselves and the bubblers. Eventually they evolve into the chloroplasts of plants.

Sometimes motorized breathers stick themselves to the outsides of bubblers, probably initially to suck their molecules as predators, later learning to drive them to new food sources for shared benefit. Eventually, some attach to bubblers already containing bluegreens, and in pushing them about discover they can take them into areas with more light. There the bluegreens find it easier to make food for the whole enterprise. This is an especially good arrangement when food is at its scarcest.

All kinds of mixing and matching occur in

79

this lengthy experimental process, always interwoven with type and extent of food supplies as well as other environmental circumstances. Complex patterns of mergers play themselves out between competition and cooperation, probably much as we see such patterns playing out today in the corporate world. Certainly it is interesting to think about the parallels as big companies swallow smaller ones that contribute new technologies, while others split off departments as independents growing and merging with yet different partners.

Some of the new cooperatives evolve cilia—"rowing hairs"—shorter and stiffer than flagellae. Arranged in rows, their rotating movements can be timed like the oars of ancient ships. Cilia are so successful that almost every plant and animal cell living today has some part or parts that evolved from these ancient rowing hairs. Bacterial flagellae also evolve into long whiplash tails. Both

are undulopodia, literally meaning "waving legs."

As the various types of bacteria merge into partnerships within common cell walls, they need to coordinate their functions more smoothly. Some of their DNA, and perhaps new incoming genes, are stored in a central information and coordination center we call the nucleus. With this innovation, the new cells can properly be called eukaryotes in their meaning "with kernel or nucleus." While the eukaryotes of multicelled creatures routinely have only a single nucleus each, the eukaryotes evolving into protists may have varying numbers of nuclei.

In sharing nuclear DNA, the various prokaryote bubblers, bluegreens, and breathers composing eukaryotes are no longer independent beings, free to come and go. They have traded their independence for the benefits of community life. Their specializations in this situation make them even more interdependent than they are in

**1900** MILLION YEARS AGO

GREAT MORTALITIES FOR THE IMMORTALS

Bacteria have no specified lifespan; they suffer no "programmed" death. When environmental factors are right, bacteria are immortal. These tiny organisms can be killed, of course, by predators, through starvation, and by encounters with kitchen-counter sprays, chlorinated water and terrorist-like antibiotics. | The light-eating cyanobacteria start an oxygen revolution. Due to their waste, the concen- tration of oxygen in the atmosphere jumps from virtually nothing to one part in five. For those masses of fermenters with no protective hideaway, an oxygen catastrophe results. A guess is that up to 90 percent of anaerobes die in the revolution.

80

81

*A moment of silence, please.*

slime cities, where they discovered they could produce food for one another and combine efforts in building shared infrastructures. In the new cells, with their increasing complexity, they are more efficiently, effectively, and tightly bound to each other than ever.

## DYNAMIC CHARACTERISTICS OF THE NEW EUKARYOTES

DNA trading and donating apparently continues both within the big new eukaryotes and among bacteria as they continue experimenting with mergers and trade arrangements. Eventually nuclear DNA includes huge collections of gene blueprints for rebuilding cells, as well as unused genes to draw on when innovations are called for. This precious DNA is protected by its own nuclear membrane.

Within that membrane, the new cells devise

almost unimaginably intricate patterns of looping and folding DNA around its protein partners. It is done in such a way that any gene can be easily found and accessed when needed. We are still trying to understand this complex storage system and its use. Whenever a nucleated cell divides, the entire complement of DNA has to be unraveled, separated, and copied. Then the two double-helical strands resulting must be relooped and folded again and again to form pairs of dense packets we call chromosomes. Each chromosome is then attached to its duplicate partner by a single tiny button organelle called a centromere until the actual split into daughter cells occurs. All is extremely carefully organized.

Eukaryotes evolve internal urban lives far more complex than those of slime cities. In addition to solar, hydrogen, oxygen, and other chemical energy engineering, electric motors, flagellae, cilia, and other forms of locomotion, further

82

## 1800 MILLION YEARS AGO
### THE EARTH RUSTS

Major atmospheric changes radically transform Earth's surface. "Red Beds," those huge, rusty piles of uniformly oxidized iron mineral, form everywhere on the planet. BIFs stop accumulating. | The

lucky among the anaerobic bacteria find mud flats and airless niches in which to survive. These relics of the Archean atmosphere still thrive in the 20th century—at the sulfurous-smelling edges

of the sea, in swamps, inside insects, and inside us. | In oxygen-tolerant, and now oxygen-loving, bacteria, grand innovations continue. Their mode of respiration is about to lead to new life-forms

emerging from a dramatic symbiogenesis.

Over deep time, anaerobic life-forms not only find special niches, they make them. This beautiful "feather" is actually a specialized organ in the intestine of beetle larvae, symbiotically made by and for glowing methanogenic bacteria.

technologies are invented. Gas bubbles called vacuoles control floating and sinking. Intracellular transport systems and architecture are built of elaborate microtubule structures through which supplies flow and which give cells remarkably strong yet flexible forms. Until recently, we did not even know that such cell skeletons— cytoskeletons—exist, because the chemicals with which we prepared cells for study dissolved them. We had assumed cells to be floppy bags of organelles and a nucleus floating around in liquid cytoplasm.

Donald Ingber at Harvard and M.I.T. now shows how these internal cellular skeletons self-assemble and function in the lives of cells. He draws on the work of scientific architect R. Buckminster Fuller, best known for his geodesic domes, to interpret and model the cytoskeletons visible in microphotographs as geodesic "*tensegrity*" structures.[35] Tensegrity, from the words tension

and integrity, refers to self-stabilizing structures in which counteracting forces of compression and tension are balanced. This gives their forms dynamic shape and resilient strength. Our bodies are tensegrity structures, raising us off the ground against gravity by the tensegrity of a muscle, tendon, ligament, and bone system.

To see what goes on in cells, Ingber builds Bucky Fuller–like models such as compressible geodesic spheres made of rigid hollow straws held together by elastic threads running through them, which snap back into spherical form after compression. He also studies sculptor Kenneth Snelson's tensegrity structures as models for internal cellular architecture. Snelson uses hollow tubes as struts joined by cables such that the tubes do not touch but are held in formation by the cables. His long sculptures seem to float on air as they extend horizontally outward from their bases like see-through giraffe necks or rise

84

## 1700 MILLION YEARS AGO

### METAMORPHIC MERGERS

In many places on Earth, complex new cells arise as microbes permanently merge. These community members form consortia in new ways: some eat but do not totally digest other microbes; others form peaceful alliances. Protoctists, whose name means simply "first established beings," are like mythic chimera, assembled from distinctly different beings. Sudden symbiotic alliances, supplemented by gradual mutations, transform the face of life. The great Kingdom Protoctista is born, launching life toward increasing complexity, new perils, and new potential.

*Alga happily photosynthesizing and leaking foodstuffs
to other less fortunate beings in its surroundings.*

vertically in apparent defiance of gravity.

Charting the various structural components of tubules and filaments in cells, Ingber has been able to show that they work as a sophisticated system for moving the cell from place to place or for changing its entire shape dramatically and optimally for different circumstances, such as stress. Tensegrity geodesics have been noted in chemistry by naming spherical groups of carbon atoms buckminsterfullerenes or "buckyballs." Ingber recognizes the parallel tensegrity structures of cells and their component molecules, such as intricately folding and refolding DNA and proteins. The study of protein folding is a whole area of microbiology in itself.

Videos of living cells show what Ingber calls these "gossamer contractile microfilament" structures in continual motion, growth, and proliferation to accommodate and meet systemic needs. He also observes filaments and tubules connecting

chromosomes to each other and connecting the nucleus directly to the cell's outer surface.[36, 37] Almost certainly this research will continue to contribute to our understanding of both DNA and protein as self-organizing materials that can respond intelligently to life's circumstances.

Inside eukaryotic city cells, then, is constant dynamic movement of the symbiont inhabitants and metabolic processes: DNA opening, copying, and refolding; RNA molecules shuttling to and from DNA in the making of new proteins; the traffic of enzymes assisting these and other chemical reactions; the flow of supplies and wastes; the ever-revised mobile architecture

---

**CONCEPT NOTE**

## WHAT ARE PROTOCTISTS?

Protoctists (the smallest called protists) are all living things other than plants, animals, fungi, and bacteria. Ubiquitous in damp, wet, watery places, these amazingly diverse beings reside everywhere from ocean abysses to ephemeral dewdrops, from moist plant tissues to dark deeps of animal bodies. Some 250,000 different species of protoctists are estimated to exist today! | The earlier joint venture of sluggish fermenting bacteria with microbial speedster spirochetes may have been a prime mover in the development of protoctists. Partners for some time, this hardy combination forms permanent attachments to a new larger cell and gets it moving. Some slide inside the cell and eventually become little organs of motility for that cell. | Some mergers also take oxygen-respiring

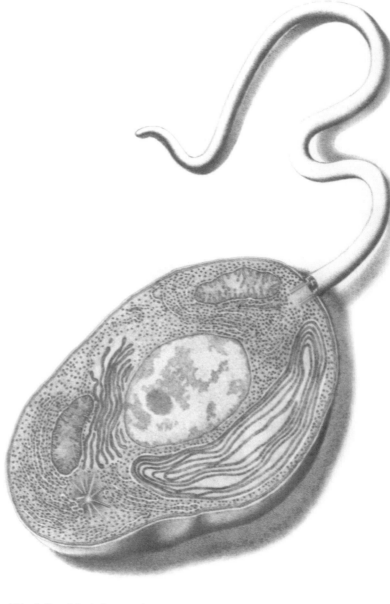

*All Earth life consists of only two kinds of cells: the prokaryote (pronounced pro-CARRY-oat), which has no nucleus, and the eukaryote (pronounced you-CARRY-oat), which does. All protoctists, animals, plants, and fungi are eukaryotes. The two cell classifications mark a great divide in all Earth life. Bacteria founded the eukaryotes through a symbiotic merger, creating the great chasm.*

bacteria as partners. These consortia set the stage for two more great Kingdoms of life: Animals and Fungi. Not about to be left out, cyanobacteria join the fray and convert many protists to photosynthesis.

This distinguished clan, the algae, expands, eventually giving rise to the great Kingdom of Plants.

*Oxygen-energized protoctists wildly diversify life. Planktic (floating) and benthic (bottom-dwelling) beings thrive, thanks to their symbiotically acquired, air-breathing mitochondria.*

## 1 6 0 0  MILLION YEARS AGO

### THRIVING IN AN OXYGEN WORLD *The Mighty Mitochondria*

Mitochondria reside, sometimes by the hundreds, inside each of our cells. They respire the oxygen that keeps alive the cells of all animals, plants, fungi, and most protoctists. | Mitochondria look like the free-living symbiotic bacteria from which they came. They do their own thing: they have their own private DNA and they grow and divide on their own inside each cell. Fortunately for all oxygen-breathing organisms, mitochondria cannot abandon us, as they can no longer live outside of our cells. | Have you ever thought of yourself as akin to a mitochondrion, living within the protective cell of our Earth? What part do we play in this symbiotic planet?

itself; and the continual molecular traffic in and out of cell walls.

The walls themselves have special security systems that identify, admit, and release only certain bacteria and free-floating molecules, sometimes letting through much larger ones than they exclude. How the cell wall communicates with the nuclear DNA-protein partnership and the rest of the cell is not fully known, though microtubule channels for such communications are newly known through Ingber's work.

Undulipodia and cilia, both based on microtubule tensegrity structures, are fixed into these walls in all sorts of patterns. The overall shapes of the hundreds of thousands of species of eukaryotes living as single-cell creatures today vary enormously.

Thus, pioneering eukaryotes evolve into full-fledged protists, the life-forms from some of which all multicelled creatures—animals, plants, and fungi—evolve. But their complex inner organization demands that they work out a different process of reproduction than that of their bacterial forebears and symbionts.

## Sex and the Single-Cell Creature: Procreation Meets Co-creation

Recall that bacteria reproduce by mitosis—splitting or budding into two "offspring"—and engage in sex by fusing with each other long enough to exchange some DNA. Recall also that sex is defined as the combining of DNA from more than one being in a single individual, so if one bacterium gets a bit of DNA from another, they have engaged in sex. Reproduction and sex are thus separate and unrelated activities in the world of prokaryotes. We could think of these two activities as fission and fusion, or procreation and co-creation. The co-creative sexual activity of

89

bacteria around the world is what we have called their World Wide Web of information exchange.

The marvelous sexual freedom of DNA-trading bacteria permits them to keep tiny streamlined bodies that are ever changeable by translating newly acquired DNA information into new protein-based systems. This gives them tremendous flexibility in their lifestyles as well as great resilience under adverse circumstances. We humans speed up their trading system today by flying from one part of the world to another, carrying myriad bacteria with us, leaving many of them wherever we stop, and picking up new microbial "passengers" in turn. That is why their immunity to our antibiotics, for example, spreads so quickly around the world—if one bacterium acquires the genes giving it immunity and spreads the genetic information locally, only one need be airlifted from that location to start spreading it elsewhere.

The bacterial style of trading DNA a few genes at a time is less common in eukaryotes because their DNA is packaged into intricately folded chromosomes within nuclei and is thus less readily available for free trade in bits and pieces. But eukaryotes retain the ability to fuse temporarily or permanently with each other, and they experiment a good deal in the trade of entire nuclei. These may be assimilated into the host nucleus or kept as one or more additional nuclei. Some protists evolved to have multiple nuclei as a species characteristic—Giardia, for example, one of the few protist species to cause us trouble, always has two nuclei.

The ability to assimilate complete extra sets of chromosomes from other eukaryotes into a single nucleus is important in the evolution of the link between sex and reproduction. Once again the story seems to begin in competitive crisis and end in cooperative solution.

Crises such as extreme temperatures, lack of moisture, or food shortages drive our protist ancestors to resort to cannibalism for the sake of survival. Sometimes, when they do not digest their own kind entirely, they assimilate the victims' DNA into their own, thus doubling their chromosomes. The fusion of two sets of chromosomes into a single nucleus if they are from different protists—even if one protist has consumed the other—is technically a sexual union. Non-hostile fusions with similar results may also have occurred.

A doubled number of chromosomes may come about in yet another way: through interrupted mitosis. Like bacteria, all early nucleated cells reproduce by mitosis, but the first stage in that process for a eukaryote is to copy all of its chromosomes into a duplicate set. If mitosis is interrupted after that stage but before the cell divides, the doubled number of its chromosomes may be kept in its nucleus.

Multiple complements of DNA may be useful in poor conditions, as they make a wider array of genes available, but they may also become burdensome when conditions become favorable again. For some such reason, protists learned not only how to acquire extra sets of chromosomes but how to shed them. The process of halving a cell's chromosomes is called meiosis, which means "lessening." According to Margulis, who traced the story of sex as well as the story of cell symbiosis, some early protists seem to become experts at doubling and halving their chromosomes. Probably they did so in accord with the demands of changing conditions from drought to plenty and back.[16, 38]

In any case, a double set of chromosomes in a single nucleus eventually becomes the normal complement for those protists that will evolve into animals and plants. Their mitosis occurs as follows: nuclear DNA is uncoiled and copied, then packed into double the original number of chromosomes and buttoned together in duplicate pairs as described earlier. The nuclear membrane then dissolves to release them, and the cell constructs a special tensegrity structure of microtubules called the mitotic spindle on which the pairs of doubled chromosomes line up. They unbutton themselves from each other and attach to the spindle tubules. The two members of each pair are then pulled in opposite directions down the spindle to two ends of the parent cell so that each end gets the cell's original number of chromosomes. The cell wall pinches inward until two daughter cells are formed and they separate, each with its full complement of chromosomes within a newly forming nuclear membrane.

This is the only way nucleated cells reproduce until cannibalism, nuclear fusion, and the evolving talent for doubling and halving whole complements of chromosomes combine with accidents of timing to work out a functional system of integrated sexual reproduction. In sexual reproduction, two parent cells fuse into one offspring. And here meiosis becomes essential. If there were no way of halving chromosomes prior to fusion, each offspring would get a full set of chromosomes from each parent—together making twice the normal number. Their offspring would have four times the normal number, and so on, to suicide.

Sexual reproduction thus requires that parent cells halve their number of chromosomes by meiosis. In sexually reproducing life-forms, the normal complement of chromosomes is *diploid*— a set of matching pairs acquired from two parents. Before reproduction, each parent must form gametes—specialized sex cells created by meiosis—thus reducing the diploid number of chromosomes to a haploid, or halved, number. Two gametes from two parents can then fuse into one offspring that has a normal diploid number of chromosomes.

Thus sexually reproducing protists fuse DNA from two parents within the process of reproduction, a system passed on later to multicelled creatures, whose gametes are, for example, pollen, eggs, and sperm. In sexual reproduction, the chromosomal DNA from the two parents must match gene for gene, though there may be different versions of the same gene, called alleles, as in blue or brown eye color. When mismatches become significant, offspring may be sterile, as happens when a horse and a donkey are mated to make a mule. With greater mismatches, reproduction becomes impossible, thus separating creatures into different species.

Sexual reproduction took a long time to evolve, and there have been many intermediate stages in its working out. Protists to this day maintain the alternative of mitosis, some of them still using it exclusively, others alternating between mitosis and sexual reproduction. The tiny slipper-shaped *paramecia* most schoolchildren have seen in a drop of pond water under a microscope, rowed about by their spinning cilia, have their own unique sexual habits. Like many protist ciliates, they can reproduce their nuclei and trade the extras with a variety of "sexes," or mating types, much as bacteria trade genes. But they are not at a loss if no partner can be found —they simply resort to fusing their extra nuclei with each other in a process called autogamy.

Most algae—protists with photosynthesizing plastids—did eventually learn sexual reproduction. We know the biggest algae as brown, red,

---

**1500 MILLION YEARS AGO**

Sex *A Survival Strategy*

For most of our single-celled ancestors, reproduction and sex are entirely distinct. Reproduction involves making more individuals. Most Earth organisms reproduce in single-parent style: by fission, budding, or forming small internal offspring cells. Sex involves fusion of genetic material from at least two individuals. Sex evolves as a survival strategy. In times of extreme stress—colds of winter, drying summer heat—our protoctist ancestors resort to cannibalism to survive. Some do not totally digest their meal; they become doubled beings and, gobbling still others in order to survive, most bloat up and die. When the environment rebounds, the survivors need to shed their doubleness and tripleness to avoid dying. These problem solvers evolve ways to regularly double every

92

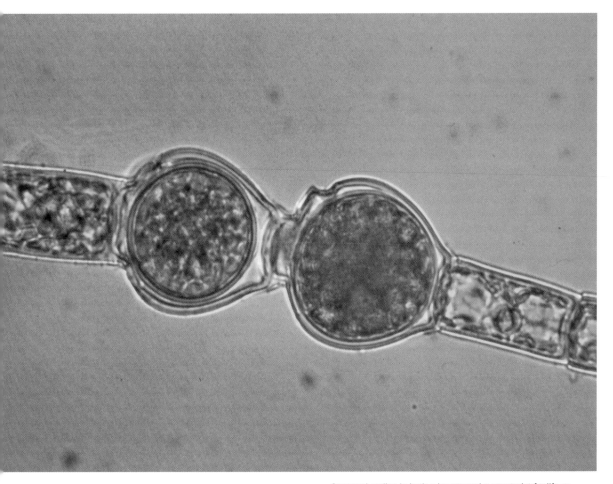

93

*Seasonal cycling is both a beauty and a constraint for life on Earth. Sex originated for survival, not for reproduction. The orange structure formed as a result of sexual doubling in this green alga.*

**winter (or dry season) through sex, and relieve the doubling every spring.**

and green seaweeds, some of which look remark-ably like plants, with roots, stems, and leaves, though they are still colonial protists, or even huge single-cell protists. Green algae are classified by two types of sexual reproduction: those form-ing gametes that swim to find each other by way of undulipodia, and those whose gametes simply fuse without developing such locomotion.

Sexually reproducing or otherwise, protist giants in a world of bacteria evolve ever new patterns and lifestyles tailored to different environments. It has taken roughly two billion years to evolve the first two kingdoms of life: bacteria, also called monera, and protists. For most of another billion years they compete over, share, and alter the world's resources.

94

## MIXING, MATCHING, AND MOVING MINERALS

Protists are prime examples of nature's mixing and matching schemes. Not only are they composed of various former bacteria as symbionts, they now take each other aboard for a wild variety of mix-ing and matching lifestyles. We have given up try-ing to speciate bacteria, and protists do not make the work of the cataloguing microbiologist much easier. Nevertheless, hundreds of thousands of species of protists have been identified.

The most visible of them are the colonial "seaweed" algae, which account for a great deal of the photosynthesis at the base of food chains. Many algae can encapsulate themselves as cysts with hard shells to survive long periods of drought. Some ciliates, such as paramecia, take in individual green algae to make food for them by photosynthesis. This habit is later adopted by

---

CONCEPT NOTE

WHY DO SOME CELLS DIE? *Shaping Up*

Complex multicellular colonies of protoctists form. In many successful communities cells become specialists. A different kind of dedication to one's community prevails as the "superorganism" (the new larger individual) evolves. Colonies elaborate on techniques developed for seasonal relief of doubling and invent organized "celli-cide" as the super-organism grows and increases in complexity.

Eventually, exquisitely organized individuals, hun-dreds of millions of cells working together, emerge. The hundred trillion cells in the human body shape themselves through differ-entiation and selective death. Programmed death of certain cells is required for differentiation. Think of a block of marble. Without Michelangelo chipping away just the right bits, no figure of David ever appears. Without death on

*How do we allow the fear of death to shape our lives?*

95

cue, no embryo, brain, or immune system develops. Scientists call such programmed death "apoptosis," a Greek word meaning "the falling away of petals from a flower."

so-called green clams, who also keep them for that purpose. There are more than eight thousand species of ciliates alone. Some have developed fancy weaponry, such as long, flexible, coiled tubes with barbed, poison-laden tips. These can be deployed suddenly by injecting water into the tubes, sending them swiftly out like harpoons to poison and spear prey, bringing it back as the tubes are retracted. These amazing ciliates are often "adopted" later in evolution by corals and anemones, which first devour and then keep them as "stinging cells" we call *nematocysts*, to assist in capturing food.

In the course of time, protists make themselves as indispensable to Earth life as did the free-living bacteria that collaborated in forming them. Most ocean plankton—microscopic creatures living in vast numbers at the surface—are protists that continually contribute new nitrogen, oxygen, methane, and other emitted gases to both ocean waters and the atmosphere. Sulfur dioxide gas produced by plankton actually seeds the water droplets, forming clouds, as mentioned earlier, thereby significantly affecting the cloud cover over much of the planet. Because cloud cover reflects light and heat back to the atmosphere, plankton play an important role in the planet's natural warming and cooling system. Dead plankton settle to the ocean's bottom in vast numbers, contributing their share to the formation of continental shelves and seafloor sediments.

Later in evolution, around the time of the dinosaurs, the most spectacular effects of protists are accomplished by those most spectacular in themselves. They are highly prolific protists that build themselves shells of the most intricate geometrical designs, moving mountains of minerals as a result and even regulating the oceans' salinity. Among them are chalk-making haptoprotists, diatoms, radiolaria, and relatively huge foraminifera.

---

## 1400 MILLION YEARS AGO
### NICHE MASTERS *Land Ho!*

Bacteria move inland, still testing their breathy expertise in new niches. They quickly settle into rivers, ponds, and the soils which erosion processes are rapidly creating. | A few ambitious microbes advance to a rough frontier. Expanses of desert become crust-covered communities. Although rare today, these ancient communities still bind the grains of desert sands. | Just as their early bacterial ancestors loved hot springs and acid, some microbes spread in icy climes. The particularly rugged outdoor types take to bare rock and mountain heights, with a breadth of metabolic strategies to meet their needs.

*These Norwegian cliffs teem with cyanobacteria that are virtually identical to their rugged fossilized ancestors.*

98

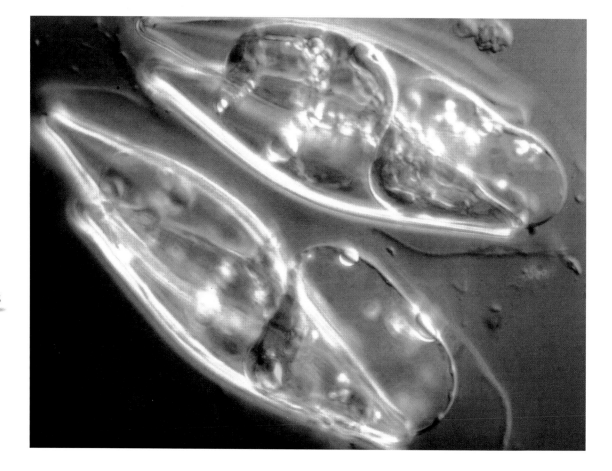

## 1300 MILLION YEARS AGO
### CORPORAL MERGERS

Protoctists have co-evolved with photosynthetic bacteria for some time. Swallowed up but not digested, some bacteria provide the protoctist with a constant supply of food, which makes them be self-sufficient algae. The engulfed bacteria receive a safe, comfortable home and rapid transport to sunlight. With their endosymbiotic (living inside) bacteria, the protoctists, now algae, are virtual living greenhouses. These cross-species associations celebrate success with permanent mergers. The bacteria transform into new organelles called plastids. Eventually, splashes of green and red seaweed will decorate Earth's coasts.

The first two are classed as algae, while the second two are assemblies of themselves and other beings they eat and adopt within, including diatoms, softer algae, and wheeled protists called sea whirlers.[18]

Among the chalk makers, the most prolific is Emiliana Huxleyi, fondly called Emily, who makes her ornate calcium carbonate shell as a sphere of daisy shapes, each produced in two hours flat. Some chalk makers build several layers of such fancy spheres of complex circles within the outer spheres for no apparent reason other than exuberant self-expression. In the process of exercising their artistic abilities, countless billions of plankton algae Emilies pull huge quantities of $CO_2$ from the

atmosphere, and pump back equally large quantities of gas that becomes sulfuric acid in sunlight and seeds clouds. Having completed their aesthetic expression and their climate control missions, their lovely little daisy balls sink to cover huge expanses of the ocean floor. The bodies of chalk makers literally build the famous white cliffs of Dover on the English coast as ancient deposits of their shells are pushed out of the sea by the endless motions of Earth's crust.

Diatoms play a major role in the cycle of rock dissolved into salts and minerals by wind and rain, then carried by streams and rivers to the sea. Life in the sea needs these building materials, but like the gases of the air, they must be kept in balance. Too many of them

99

would choke off the oceans' creatures.

In the course of evolution, vast quantities of minerals are carried to the seas until they become supersaturated. But somehow the right plankton evolve to balance things out. Hundreds of millions of tons of silica (silicon dioxide) alone, for example, are washed into the oceans every year. Huge swarms of microscopic diatoms use these silica supplies to build fancy pillbox–shaped shells sprinkled with tiny holes in intricate patterns. When they die, the diatoms sink to the bottom, leaving the silica shells to settle into rock—300 million tons of new silica rock every year. The number of diatoms in the sea naturally adjusts to the supply of silica brought by the rivers.

By the time diatoms appear, later than Emilies, the seas are supersaturated with silica. In time, however, enormous numbers of diatoms come to maintain a very precise balance between incoming silica and silica sediment deposits.

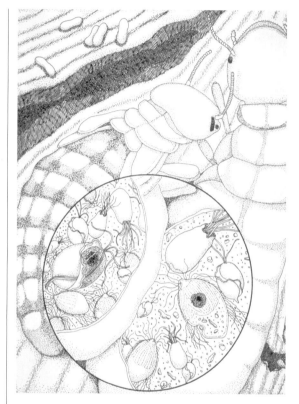

*The microbial world inside a New England termite* (Reticulitermes flavipes) *is seen in this circle of microscopic light.*

## CONCEPT NOTE
### WHAT IS AN INDIVIDUAL?

Identity is a process, not an object. All Earth life is connected through a common ancestry. Each "individual" (each organism)— cow, beetle, daisy, human —is actually a consortium of transformed and still-living other beings. *Mixotricha paradoxa* ("paradoxically mixed-up hairs"), as seen in the termite community, may help to explain the fractal, nested-network nature of life. A termite nest functions as a superorganism: each nest is an "individual" made up of thousands of termites with specialized, integrated roles. Within an "individual" termite are wall-to-wall microorganisms numbering up to $10^{12}$ (a trillion) bacteria and $10^7$ (10 million) protists. A termite's hindgut microbial community (an anoxic habitat for successors of ancient microbes) helps digest the wood consumed by the

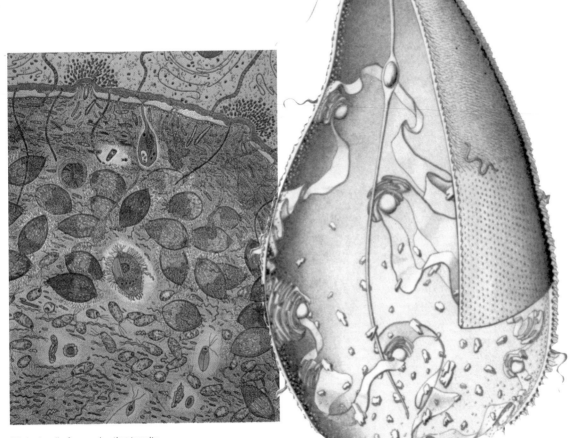

Hindgut wall of a wood-eating termite.

101

chewing machine. | Within that hindgut microbial community lives a beautiful tiny protoctist called *Mixotricha*. It is actually a consortium of populations: one nucleated cell, two kinds of spirochete bacteria, a rod bacterium on the surface, and internal (endosymbiotic) bacteria. *Mixotricha* is in the process of emerging a new "individual."

*Bosch? Dali? No, it's* Mixotricha paradoxa. *From 250,000 to 500,000 tiny spirochetes move the "giant"* Mixotricha *through the viscous habitat.*

Once again the self-organization processes of Earth are evident.

Some diatoms solve another problem in an unusual way worth looking at. Diatoms normally reproduce by mitotic splitting. But this causes a problem, because the "lid" of their rigid pillbox shells is larger than the box they cover. In splitting, half their offspring get the smaller box bottoms and half get the lids. Each makes a new matching part to complete its box. Unfortunately, these diatoms have only learned how to make box bottoms in response to inheriting either part. While this is fine for the lineage that inherits lids just like the original ones, those inheriting bottoms must use them as tops and make smaller boxes to fit into them, meaning each succeeding generation gets smaller.

Amazingly, they solve this problem by sex. At some danger limit, the line of ever-smaller diatoms ceases to reproduce by mitosis and engages in meiosis, producing gametes with haploid chromosomes and no shells. These gametes fuse with each other to produce new diatoms that apparently have full DNA plans for both box tops and bottoms of normal size. Even the smallest diatoms can apparently bring their lineage back to full size in this way.

Other marine protoctists, such as radiolaria and foraminifera, build yet more spectacularly ornamental shells from opaline silica. They look for all the world like crowns, Japanese warrior helmets, and delicate Christmas tree ornaments. They also invent mobile spikes that protrude through the lacy openings of their shells and wave to swim or "walk" them about. Under a microscope we can see their bases as complex hexagonal arrays of microtubules in tensegrity structures that account for their motion. We can see similar designs in much larger starfish and sea urchins. Ancient *foraminifera*, though single-cell protoctists, grow as large as ten centimeters in diameter, housing a variety of other symbiont protists within.

Limestone formations all over Earth owe their existence to these various ancient shelled protists. Once ocean sediments, they are later pushed up from the seas and remain treasure troves of these tiny fossils. While many mineral-shelled protoctists do not evolve until well into the age of dinosaurs, earlier forms are already instrumental in balancing the chemistry and temperature of the atmosphere and the seas, while transforming Earth's crustal materials from eroded rock to living matter and back to rock.

## FROM MULTICREATURED CELLS TO MULTICELLED CREATURES

Just as the first protists were multicreatured cells formed by cooperative teams of bacteria, the first multicelled creatures are formed as colonial cooperatives of protists.

Protists begin living together in colonies by sticking together after reproducing rather than floating or swimming off on their own. In some cases, each protist in the colony is as independent as if it had gone its own way. In others, the protists invent new ways to communicate, such as sending chemical messages among themselves.

Chemical communications make it possible for individual protist cells to harmonize their activities with each other. They form hollow multicelled balls, for example, and coordinate their rotating undulipodia or cilia. Beating as rhythmically as the oars of our own ancient galley ships, they can move a whole colony smoothly in any direction. In some protist colonies, certain individuals become specialists at locomotion; in others, some cells can produce glue to stick their colonies to rocks. Other divisions of labor include specializing in making food by photosynthesis, or in capturing food floating nearby and digesting it to produce energy and building materials for the whole colony.

In some colonial arrangements, a few cells come to specialize in sexual reproduction. Halving their chromosomes by meiosis, they produce haploid gametes that can fuse with other gametes of the same colony or, if they are liberated, can meet and fuse with similar gametes from other colonies of the same type. Whole new colonies are formed by mitosis from these fused cells, beginning the evolutionary pathway toward the true multicelled animals that begin with embryos cloned from fertilized cells.

A fascinating step in the evolution of multicelled creatures from protists is the slime mold, a life-form in between protists and multicelled fungi. Many slime molds start out as individual amoebalike protists. These soft-bodied nucleated cells move about by extending pseudopods—literally, "fake feet"—and pulling their bodies into them. These little slime beings spread over their environment, hunting for food to engulf and digest, and in some species even reproducing sexually. But when their food supplies run out, they send out chemical messages that attract them to one another until they are gathered together, climbing on top of one another to form a visible sluglike community. Sometimes you can see such jellylike masses on the underside of rotting leaves or logs.

Other species of slime mold have many

103

reproducing nuclei scattered through a single jellylike sheet that can dry out and later reconstitute itself when moist conditions return. These types also contract into a sluglike form when food runs out. The sluglike phase of the slime mold moves itself about as a whole, seeking a place where it is out in the air to stop and sprout a stalk that forms a fruiting body at its tip. From this bulging tip it releases spores—tiny packets of DNA and protein similar to those of true molds, which are fungi. Breezes blow the spores through the air until they settle in new moist places to form new amoebalike creatures or jelly sheets, beginning the cycle all

104

over again. As with bacteria and other protists, hunger drives slime molds to this innovative and cooperative lifestyle.

Colonies of protists in the ancient seas evolve in two directions. Those that make their living primarily by photosynthesis become algae, including all the seaweeds. Through specialization of cellular function, other protist colonies evolve into the first animals, undoubtedly many soft-bodied ones that leave us no records. Fortunately, some do leave fossil imprints. Take, for example, the quite spectacular "jellyfish" and "soft corals" found in England, then more abundantly in Australia, where

## 1200 MILLION YEARS AGO
### THE ONE AND THE MANY

The explosive growth of the fossil record of algae and other organisms moves some paleobiologists to dub this the Big Bang period of eukaryotes, beings made of cells with nuclei. | Over all Earth, protoctists form communities with bacteria as well as with other protoctists. Manic mixing and matching of organisms occurs. | In "colonial" protoctists, a few cells break off from the individual body and regenerate the entire organism. This is not possible for most multicellular protoctists, whose cells, like ours, differentiate (specialize).

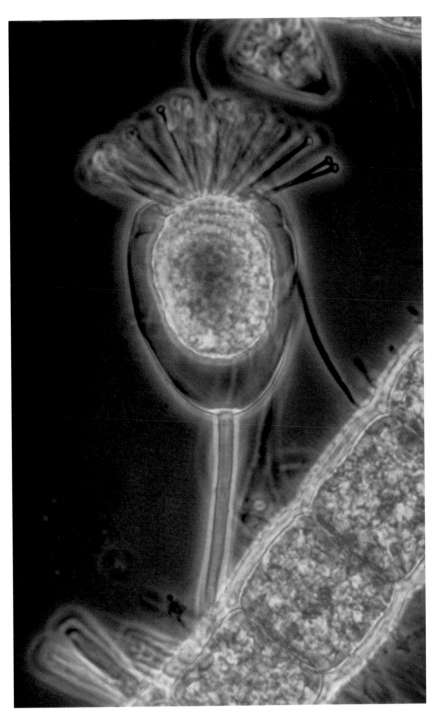

(left) The white blobs are free-loading red algae living on photosynthesizing other red algae. The white blob actually sends its nuclei into the red photosynthesizer. It then directs the transfer of photosynthetic products back "home."

(right) A filamentous alga, coated with bacteria, supports the base of a tuliplike ciliate that grabs even tinier organisms out of the water through multiple feeding tubes.

105

they became known as Ediacaran fossils, and now on every continent except South America.

The tiniest Ediacaran creatures are wormlike and smaller than pins, while the largest are shaped like huge flowers, quilted mats, banners, fringed pinwheels, or giant leaves, no few of them as much as several feet in diameter or length. Apparently slowly drifting, peaceful beings, they evolve at the mysterious boundary between multicelled colonies and genuine animals, most of them very different from later life-forms and possibly similar to what we might find on other planets evolving life.

Giving them strange names such as phyllozoon—"leaf animal"—scientists still speculate on whether their food comes from bacteria and protists absorbed from the seafloor beneath them or from a descending rain of surface plankton—or whether they house large numbers of symbionts to produce their food from within. Whatever their lifestyles, all of them are now

classified as Ediacaran, the name of a new period from 600 MYA to 540 MYA, newly recognized and tucked in between the end of the long Archean era and the Paleozoic era.[39, 40]

Between one billion years ago and the time we find early Ediacaran life, Earth's crust goes through one of several breakups of single landmasses—this one called Rodinia—into separate continents and moving some of them back together again as Gondwana. By the time the Paleozoic era ends, other continents will have fused as Laurasia, which in turn will merge with Gondwana to form yet another supercontinent called Pangaea. Such tearing apart of the crust, and the consequent mountain-raising crunches when continents again move together, change the whole face of Earth dramatically. Changing land formations in turn change the levels and flow patterns of oceans, seas, and inland waters. New weather patterns develop and as many as five ice

106

| 1100 | MILLION YEARS AGO |
| --- | --- |
| INTO THE BREACH | |

Due to the brevity of human life, we struggle to grasp the movement of continents and ocean crusts. From the perspective of deep time, we can feel Earth move under our feet. | Earth is rest-less. Major global rifting occurs in a very short amount of geologic time. Great valleys open within the continental shield, rivers pour in and new oceans form. Plate collisions fold Earth, and huge mountain chains rise. Volumes of magma spew from the deep. | Continents are the raised portions of tectonic plates.

107

*Topography is as mountainous
and deeply gorged under the
oceans as it is on land.*

ages that cover much of Earth with glaciers occur around 700 million years ago, apparently triggered by Rodinia's breakup.

The plate tectonics of continental movement are considered a leading cause of ice ages. When landmasses shift to high latitudes, especially if they are simultaneously lifted higher in altitude on the backs of the great plates, snow and ice can accumulate on them to cause ice ages during which multiple glaciations occur. Because Earth's crust changes so dramatically, the ice age associated with Rodinia's breakup is the earliest we can currently detect. We know of others around 440 MYA and then around 200 MYA when the Pangaea supercontinent was located in the Southern Hemisphere all the way to the South Pole.

Note that continents presently predominate in the Northern Hemisphere and that we are in the midst of an ice age beginning three million years ago, with glaciations in cycles of about 100,000 years—90,000 years of cold interspersed with 10,000-year interludes of warmth. Other factors in ice-age production seem to be related to Earth's position and tilt cycles in relation to the Sun and to atmospheric $CO_2$ levels, though the latter seems doubtful in early glaciations.

Ediacaran organisms thus seem to be associated with the warmer waters of post–ice age times, when the land is still barren of life except for bacterial colonies lending their purple, red, yellow, and green color to an otherwise drab landscape.

## ANIMALS AT LAST

The Paleozoic—"ancient animal"—era officially begins with the so-called Cambrian Explosion, 540 MYA, when animal fossils suddenly appear clearly and in abundance. Though more than

108

| 1 0 0 0 | M I L L I O N   Y E A R S   A G O |
MINI-MINERAL MARVELS

As geological processes create and metamorphose Earth rocks and minerals, life radiates (multiple species diverge from common ancestors). New protoctist species evolve, each with their own special mineral interests. Some of these microbes produce their own minerals (biomineralization), using them in diverse and creative ways. Bacteria learned long ago, for example, to make magnetite, which they use for compass orientation in muds and shallows. Although still in early development, the protoctists display manufactured minerals in a variety of styles unmatched by other kingdoms of life.

Scientists today recognize over 60 different inorganic minerals produced by life. A variety of organisms, ranging from bacteria to humans, participate in the production processes. This living protist is a foram. The foram makes its shell of calcium carbonate. Diatoms, smaller symbiotic protists living inside the foram, make their own silica shells.

109

Many contemporary anemones are symbionts with hermit crabs: the crab provides a free ride, the stinging anemone provides protection. Crabs move out and on when they grow too large for their borrowed shell homes, rather disruptive and discomforting for the anemone. This beautiful golden shell is actually a biomineralized overlay and addition created by an anemone for its crab partner. As a crab grows in size, its anemone partner can add to the shell to maintain a fit that is just right.

three billion years—over 80 percent—of life's evolutionary journey was completed by then, little has been known about Archean life until recently. While our "evolutionary tree" charts acknowledged microbes as the vaguely known roots of larger life-forms, we focused on the huge branches devoted to animals, plants, and fungi. Small wonder that our attention was focused on

animal family lineages set against changing background habitats over time.

In Darwinian tradition, we assumed until recently that evolutionary change happens steadily and gradually over time due to a steady stream of accidental genetic variations that produces differential "fitness" in the members of individual species. This fitness is determined by their

*This contemporary alga is an example of one of the largest protoctists.*

*This thin rock section displays fossils 900 million years old.*

110

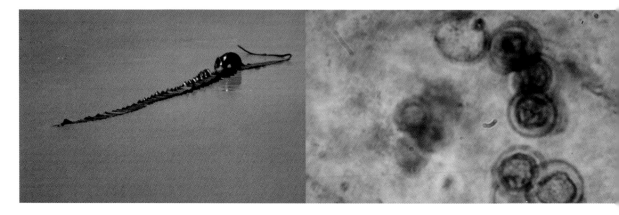

900 **MILLION YEARS AGO**

CATEGORY QUESTIONS

Abundant acritarch fossils, some giant-sized by micro-standards, are thought to be fossil cysts or other life stages of algae. Although we cannot discern exactly "who" they are, they speak to us of life's diversity over millions of years. Categories are maps, but maps are not the territory. In the game Twenty Questions, the basic categories of animal, vegetable, and mineral underestimate the richness of life-forms. Similarly, basic categories such as "small," "large," "single-celled," and "multi-cellular" can confuse us.

environments and leads to differential survival rates—a process called natural selection. Thus, we eagerly set out to trace the lineages of our most beloved animals—so eagerly that museums arranged fossils in mistaken sequences based solely on appearance, as in the case of the later withdrawn eohippus—"dawn horse"—lineage at the American Museum of Natural History. In fact,

clear lineages are extraordinarily difficult to find when we try to go far back in the fossil record. We are still working, for example, on the lineage of the largest creatures—whales.

As we learned more, evolutionary theory, especially through the work of Stephen Jay Gould, shifted to ideas of punctuated equilibrium—sudden phases of rapid evolution inter-

111

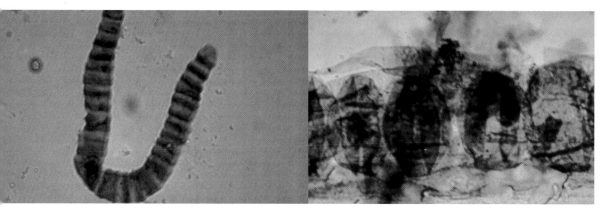

Photos of ancient filamentous microfossils; the one above and the one to the right are fossil cyanobacteria.

112

spersed with slower phases of little change.[41, 42, 43] We discovered that some species, such as cockroaches and sharks, are like bicycles in the jet age—not changing despite great changes going on all around them—while others, including our own species, change with astonishing evolutionary rapidity.

With new revelations made possible in microbiology by ever more sophisticated microscopy, we can now literally see the genes responsible for change, and are close to watching them in live action over generations of change. Again, our picture of evolution changes dramatically. No longer are genes passive molecular structures subject to change only by accident; they are active living structures repairing such accidents and traded about within nuclei, among cells, and even among species, depending on the needs of the colonies or creatures in which they play such central roles.

## 800 MILLION YEARS AGO
### SOME LIKE IT COOL

One of a world-wide series of ice ages sets in. Thick ice sheets expand over vast areas of Earth. What causes these ice ages is not completely understood. One theory attributes the ice ages to fluctuations in Earth's orbit, which affect the delivery of solar radiation to Earth. | Temperature swings can be amplified by other factors. If solar radiation decreases, for example, Earth cools and glaciers expand. Since glaciers are reflective, the expansion reduces the amount of heat absorbed by Earth. Thus, temperature spirals down. | The life system of Earth is closely coupled to climate change. Microbes have, since the origin of life, played a major role in the carbon cycle. Ocean-floor and ice-core drilling shows precision synchrony between climate change and atmospheric carbon dioxide. This coupling

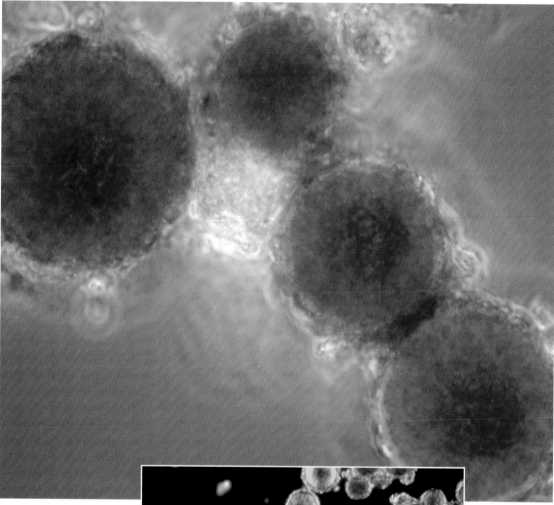

113

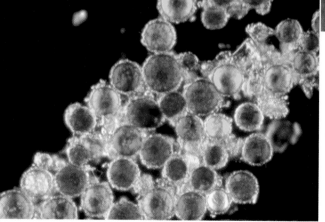

**affects the magnitude of climate change.**

*Diverse new planktic and benthic communities evolve from these ice ages, thanks to the rich nutrients found in colder waters. Some land protoctists, like these red snow algae, also like it cool.*

On the macrolevel, the view of Earth from space, looking so alive, inspired a great deal of research based on holistic systems theories, the best known of which is Lovelock's Gaia hypothesis. [3, 4] More broadly among evolutionists, whole ecosystems, and even Earth in its entirety, are now seen to evolve as entities in their own right. Biology is increasingly difficult to separate from geology as lithosphere, hydrosphere, atmosphere, and biosphere become seen as an integrated dynamic whole changing over time through self-organization processes.

In this new light, let us take a fresh look at the rise of the animal kingdom from the time of the Cambrian Explosion. Animals are defined by the fusion of gametes that gives rise to embryonic development which includes a blastula stage. The earliest fossil embryos, as well as clearly wormlike creatures, dating back to the Ediacaran period were found in a Chinese phosphate mine in 1998. [40]

Among the early animals that evolved from protist colonies are polyps. Many living polyp species so closely resemble the fossils of ancient polyps that we can get clues to their early evolution by watching them today. Anemones are polyps, and coral forests are huge polyp colonies.

An individual polyp animal is shaped like a tube, with a flowerlike circle of tentacles at one end around its mouth. The other end of the tube is stuck to a rock or to the body of another polyp in its colony. If you watch anemones in a seawater tank over time, you will see that some stay put while others wander about slowly without ever detaching themselves. Polyps catch prey with their tentacles, which they use to stuff the food into their mouths. Many types have also adopted harpooning nematocysts—literally, "thread bags" —along their tentacles to tangle their victims and paralyze them with poison barbs.

Polyps reproduce by budding. In some species, the task of reproduction is assigned to certain members of a polyp colony, which are fed by the others so they can concentrate on their important work. Polyp buds may grow up stuck to the parent, though others do something much more interesting. The newly budded polyp breaks off, flips over so its tentacles hang down, and floats off into the sea. As it grows larger, it may become a glassy bell or umbrella with a softly fringed edge of trailing streamers—a jellyfish, as we call it, though it is not a fish at all. Its proper name is medusa, a name taken from the ancient Greek myth of a woman who had snakes on her head instead of hair.

Medusae are a much more adventurous stage of polyp life that learned to reproduce sexually. Some species tried having both sexes in the same individual—as flowers and earthworms have them—while other species began making separate males and females. In any case, all medusae produce female eggs and male sperm, which fuse to make baby medusae. The baby medusa is so different from its parents that it, too, gets its own name. We call it a planula. The planula is a long, flattish blob that rows itself about freely for awhile with its fringe of cilia. Eventually, it settles onto a rock and sticks itself tight to grow into a polyp, thus completing its life cycle.

This life cycle is thus a kind of serial metamorphosis—changing form from polyp to medusa to planula to polyp. Many species of polyps still abound in the seas, looking much like their ancient forebears. But somewhere in the dim past, some polyps become discontented with the prospects of a lifetime stuck on rocks and invent cyclic metamorphosis, giving freedom to their children as medusae, though their offspring must return to the rocks to grow up. But some planulae seem to protest against the change back to a sedentary life and begin skipping the polyp phase to grow straight into medusae. Later, some of their young planula offspring skip the medusa stage, reproduce themselves sexually, and go on to invent yet more adventurous lives as free-swimming creatures that eventually evolve into fish—and much later into us.[12]

This evolutionary phenomenon of branching off into new directions from a juvenile phase has a special name: neoteny, meaning "stretched youth." Neoteny is a kind of step backward when evolution reaches some adult dead end or blind alley. It occurs when the body of the grown-up stage becomes specialized to a point preventing further evolution, but the juvenile form can still be altered.

We can see such evolutionary dead ends in polyps living today that are still very like their ancient ancestors, having evolved no further. But in some ancient species, evolution takes advantage of the fact that baby planulae, which do not yet have the specialized bodies of adults, can still change. We can trace the descent of some free-swimming creatures to ancestors who are like unspecialized planula babies. Neoteny also happens much later in evolution. If you look at a chimp family, for example, you will see that humans look like the baby chimps, not like the adults. What features did we give up to become human?

## PATHWAYS OF SPECIALIZATION: IN SEARCH OF FOOD

Animals, which include polyps such as sea anemones, jellyfish, and lobsters, as well as later insects, amphibians, reptiles, birds, and mammals,

115

may actually evolve from protist colonies that lose their bluegreens or other photosynthesizing symbionts. This would mean losing the ability to produce their own food. Instead, they must go in search of food or find ways to attract it. While evolving plants can float slowly or sit in one place making a living, animals have to develop a whole range of new organs.

They need means of locomotion, of hunting and gathering, of consumption and elimination, and even of perception, to tell them what in their ever more complex world is food and what is not. Thus they evolve all sorts of equipment— from eyes and ears to feet and wings, claws, and teeth, the marvelous tensegrity structures of skeletons and muscles, digesting systems, heating and cooling systems, brains for organizing all this complexity, and hearts with blood vessel systems to keep it all going. All this tremendous complexity evolves just because they have no chloroplasts

Ophrydium versatile *is a colonial ciliate in the process of becoming what we humans call an "individual." These "green jelly ball" colonies are microbial worlds within worlds, consortia of several hundred different kinds of microbes. Ophrydium may be the closest living analogue of the large photosynthetic and chemosynthetic protoctist colonies of the quilted soft-bodied wonders of Ediacara.*

116

(above) Green jelly ball colony

(right) Lots of zooids ("little eyebrows") in the jelly ball

---

| 7 0 0 | **M I L L I O N   Y E A R S   A G O** |

SUPPLE TESTIMONIES

**Ediacaran organisms leave quite an impression, which is rare for soft-bodied biota. Beautifully bizarre, their shapes vary from leaflike to three-armed to flat to "quilted." In the shallow coastal seas of** the "Garden of Ediacara," **photosynthetic and chemosynthetic symbionts help some of these organisms grow large, while others graze on plentiful bacteria. These gelatinous creatures have no hard** parts and no predators. **Theirs was a pre-armored world.** | Ediacaran fossils **are found all over the world. Evolutionary biologists disagree about the nature of these enigmatic beings, since none of their** relations survive to tell **the tale. The delicate Ediacarans, an evolutionary experiment in life forms, go gently into that good night.**

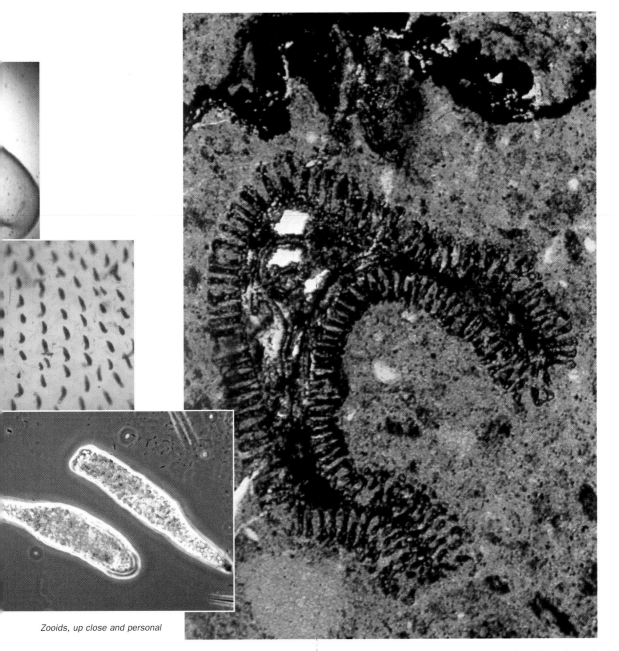

*Zooids, up close and personal*

A mystery fossil with no known relatives, the species Pteridinium *occurs abundantly on the surfaces of* Precambrian *sandstones in Namibia, southwest Africa, and other parts of the world. Some scientists consider* Pteridinium *to represent the earliest known animal fossil; others are skeptical that the organism was related to animals at all.*

and must chase after and assimilate food.

Plants, like animals, reproduce sexually. But they depend on wind, water, and animals to move their pollen and seeds about, while animals can meet and mate in their travels. Their egg-and-sperm sexual reproduction, as we said, defines them as animals, as does the division of the fertile egg into a hollow sphere of cells called a blastula early in their embryonic formation. Blastula cells form complex alliances through chemical communications, as if telling each other which ones must form what parts. Cells left out of that process, with instructions to withdraw to leave space (such as between the individual fingers of a hand), commit apoptosis, or "cellicide," for the greater good.

Ancient sea animals in the Cambrian period take one of two basic forms: soft bodies such as

118

*Beetle sperm penetrating an egg*

WHAT ARE ANIMALS?

We usually think "mammal" when we hear the word "animal." In fact, we and this Rhinoceros beetle are animals. | In the Animal Kingdom, a small swimming sperm makes it to a large egg, spurred on by its undulating tail. The fertilized egg repeatedly divides to form, in the initial stage of embryo development, a hollow sphere of cells – the animal blastula. This blastula is the defining trait of animal-hood. | The bodies of animals are individualized with special cell-to-cell connections. As the embryo cells divide, some must form alliances, while most others die on a pre-programmed cue. If these cells do not commit cellicide in the proper fashion, no animal body develops.

*The following true, or untrue, bug story cites J.B.S. Haldane, hero of evolutionary biology. At a formal dinner, Haldane was seated next to his staunch foe, the Archbishop of Canterbury. There was, of course, "polite" British exchange:*

Archbishop: *What do your studies tell you, Professor, about the nature of the Creator?*

Haldane: *He must have had an inordinate fondness for beetles.*

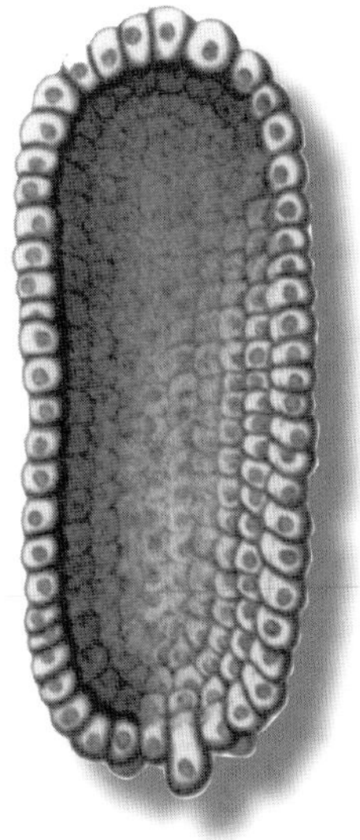

Blastula, a kind of embryo

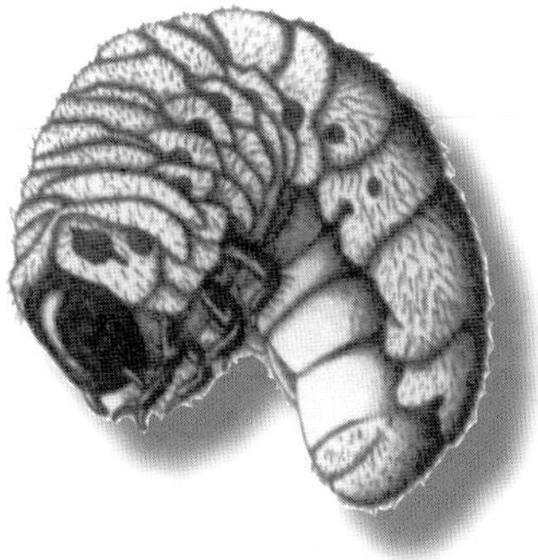

Larva

119

Beetle

*Insects innovate more successfully than do any other animals. Close to a million different species have been identified . . . and we're still counting. Opposite, beetle sperm penetrate egg, which divides to evolve sequentially into blastula, larva, and beetle.*

sponges and the anemones that look so much like flowers, or bodies armored in shells. While shelled creatures go in one evolutionary direction as arthropods—"jointed legs"—some soft-bodied creatures such as the planulae of polyps evolve into worms and later animals with backbones, such as fish.

Somewhere along the way, rowing cilia apparently turn inward to become tubes through which messages can be sent from cell to cell in early animals—the first signs of nervous systems. These new communications systems make it possible to organize ever greater numbers of cells within a single creature. Some cell groups become nerves; others evolve into stiff but flexible cartilage tubes called *notochords* running down the lengths of the nerves and protecting them. Notochords are clearly reminiscent of the tubules evolved within eukaryotes long before, again raising the question of how DNA reprograms for

120

*This beautiful little creature, just two thousandths of an inch long, is one of the smallest animals in the world. It is called a "rotifer" because the cilia of its tiny retracted wheel-like head beat and flush food toward its mouth.*

### 600 MILLION YEARS AGO
### ANIMALS ARISE

**The first animals arise when marine protoctists curtail reproduction in favor of specialization. They are very small, with only soft-body parts, so they may swim with their protoctist cousins for** **millions of years before circumstances are favorable for their preservation in the fossil record.**

Colonial coral, Alcyonian, "dead men's fingers"

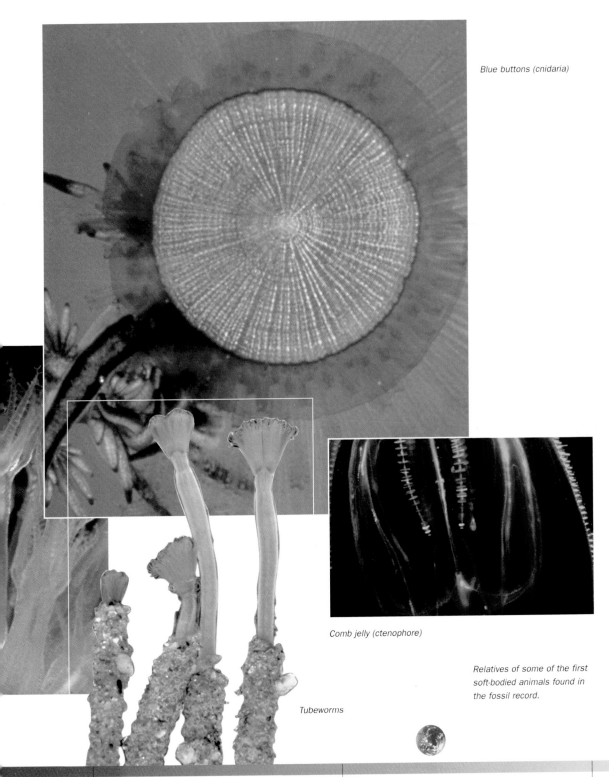

Blue buttons (cnidaria)

121

Comb jelly (ctenophore)

Relatives of some of the first
soft-bodied animals found in
the fossil record.

Tubeworms

2,000 MYA

1,000 MYA

PRESENT

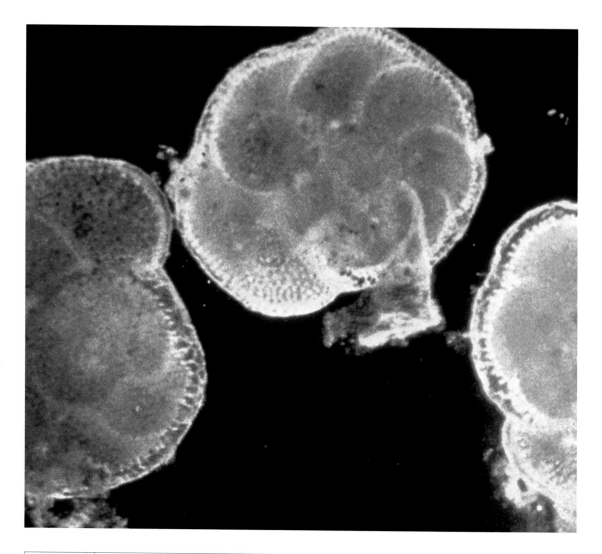

122

**580** | **MILLION YEARS AGO**

## FABULOUS FORAMS

Foraminifera ("forams" for short) are aquatic protoctists whose cells are enclosed in loose-fitting, hard, shell-like covers called tests. Found throughout the world's oceans, tests provide clues to the past. They are key biomarkers for oil companies looking for layers from the "right age" for drilling. Their presence in the desert means an ocean once covered the area. Different species of forams are very fussy in choosing their habitats; their fossils help us "read" the nature of paleoenvironments. | Foram individuals are small, but as a group they are a mighty force. The calcium carbonate of their abundant tests affects the global carbon cycle. Forams of the past unite, their tests make up the sedimentary rock of the great Egyptian pyramids and White Cliffs of Dover.

similar stuctures at larger levels of organization.

Long cells that can stretch and shrink by tensegrity attach themselves to the notochords, moving the creature in wiggling patterns and evolving later into muscles—a way of moving the creature from the inside rather than by way of external rowing cilia or motorized undulipodia.

The notochord tube wraps itself around the nerves and eventually becomes a backbone of separate vertebrae. Polyps have already evolved tentacles, mouths, and guts. These also continue to evolve into ever more complex systems. Hearts and blood vessels come into being to circulate supplies. Nerve bundles at the head end of animals become simple brains, while light-sensitive eyespots evolve into eyes, and so on.

123

*The "spots" on these living forams are their photosynthetic symbionts.*

124

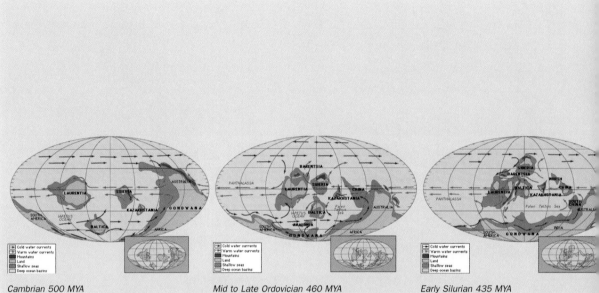

Cambrian 500 MYA

Mid to Late Ordovician 460 MYA

Early Silurian 435 MYA

## CONTINENTAL CAPERS

The Paleozoic Era runs from 541 to 245 million years ago, and geologists divide it into six major periods: Cambrian, Ordovician, Silurian, Devonian, Carboniferous, and Permian. The "Cambrian Explosion" ushered forth a great burst of life and the appearance of the first shelled animals. The Paleozoic Era ends with the greatest known mass extinction of life in Earth's history. | The Era opens dramatically: first algae and then animals make a great and perilous leap from water to land (approximately 1,000 MYA after some bacteria colonized land). These new inhabitants devise ingenious ways to carry the ocean with them. The Plant and Fungi Kingdoms make an official debut in life's drama. Earth itself changes as continents riding on their

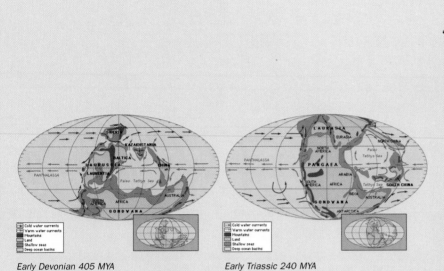

*Early Devonian 405 MYA*

*Early Triassic 240 MYA*

"California moves north at about a centimeter or two each year, producing earthquakes as the Pacific plate slides past the North American Plate. "Baja" California will move up opposite to the interior of "Alta" California. Eventually, Los Angeles will be pushed opposite Berkeley, with results we cannot predict."

Raymond Siever

125

plates congregate, converge, break apart, and recombine.

Squid and cuttlefish have the basic design of tube-shaped polyps with tentacles around their mouths, yet they are free swimming and evolve eyes as elaborate as our own. They also invent yet another mode of locomotion—the water jet. Taking water into their hollow tube bodies, their muscles squeeze it out to propel them along in the sea as suddenly as a nematocyst cell shoots water into its food-snaring poison lasso. And of course they discover the camouflage of hiding in clouds of their own ink.

In some species of early animals, mouths grow a single big sucker instead of tentacles, as in some eels, or build jaws and then teeth. Some experiment with inner skeletons; others, as we saw earlier, evolve outer shells to hold their bodies together and protect them.

Half of all animal groups called phyla originate, or blossom, during the Cambrian Explosion and many leave fossils that help us track their evolution. The greatest Cambrian fossil find is in the Western Rockies of Canada. Known as the Burgess Shale, it is unique in preserving many soft-bodied creatures that usually leave no fossils. It also reveals an ancient sea world full of wildly varied shelled animals that look like science fiction creatures from other planets—*Opabinia*, for example, with its armored shell and its long, flexible, fanged, vacuum-cleaner hose.

While mollusks such as the nautilus, barnacle, and clam keep evolving simple single and double shells, spiny sea urchins and starfish grow spiny coverings. Arthropods such as *Opabinia* go on to experiment with countless new lifestyles, including as crabs and lobsters (our favorite), and, on land, as mosquitoes and flies (our least favorite). Arthropods account for 80 percent of all animal life, including hundreds of thousands of species of beetles alone. All the while, the lineage of polyps goes on to evolve into cartilaginous sharks and bony fish.

By this time the oxygen atmosphere and cloud cover above the seas give Earth a blue-and-white look in place of the earlier reddish haze. Meanwhile, the great tectonic plate activity continues to be driven by magma flowing through seafloor rifts, while old sediments are pulled under at the plate edges, recycling crustal materials as they move the continents on in their ponderous journeys.

Ocean plankton, meanwhile, continue to absorb carbon from the $CO_2$-rich early atmosphere, leaving it buried in sediment as the tiny creatures sink to the bottom where myriad bacteria and larger creatures recycle their nutrients, some of which float back up to feed surface plankton. The abundant new animals living and dying on the seafloor and above must also be recycled. Dead multicelled creatures are excellent sources of food for microbes, and most recycling of biomass is a microbial service.

Recycling is an absolutely essential process in evolution. Just as the tectonic plates ever recycle crustal materials, so must the living biomass on the crustal surface continually recycle itself. Even our bodies are continually recycled, cell by cell, molecule by molecule, atom by atom. This is done so wonderfully inconspicuously by our physiology that we never fail to recognize ourselves, even though we are composed of entirely new molecules every seven years or so, most recycling in far less time.

The fact that death is necessary to continue populating Earth with creatures has been hard for humans to grasp and accept. But if new creatures kept coming to life without others giving up their lives and being reduced back to raw materials, the supplies in Earth's crust would soon be used up. The mass of creatures would all die together of crowding and starvation, as we are rapidly learning from our successful efforts to delay death.

Half a billion years ago (500 MYA), still near the beginning of the Paleozoic, the oceans teem with recycling life. It is time for algae and arthropods to follow bacteria onto land, where they have worked so diligently for long ages to create soils and atmospheres hospitable to larger creatures.

## ALGAE AND ARTHROPODS ASHORE: THE APPEARANCE OF PLANTS

Perhaps the evolution of plants begins on land, when algae are left high and dry between moon-pulled tides. Whether to escape predators and crowding or simply to engage in new adventures by staying ashore, they adapt to the new open-air lifestyle with great success. To survive in air, they thicken their walls, thus staying wet on the inside while the air dries them on the outside.

Fungi, a kingdom including molds, mushrooms, and yeasts, also now appear on shores, where they practice the ability to digest organic food with excreted enzymes before consuming it. This inside-out variation on digestion, along with their ability to dissolve the rock they live on by secreting acids to extract its minerals, make them wonderful recyclers. Some fungi cooperate with algae to form a single-organism partnership we call lichens. Lichens are able to live on rock, dissolving its minerals and photosynthesizing food. Lichens, like algae before them, experiment with plantlike forms.

The spores of algae and fungi blow about, developing wherever dew or rain leave enough moisture. Fungi reproduce both sexually and by way of spores and spore packets called propagules, which can lie dormant for thousands of years before springing to life when they encounter moisture again. Some fungi today are easily visible, such as puffballs and mushrooms. Others live underground, sometimes as enormous beings, such as the famous Michigan fungus that is fifteen hundred years old, covers thirty-seven acres, and is estimated to weigh over eleven tons!

Plant cells are more complex than those of either animals or fungi because they have both mitochondria and plastids (such as chloroplasts). The first real plants to evolve are mosses, often found in close association with lichens. From the time they invent a cooperative lifestyle as lichens until today, plants and fungi have a very close association. It may even be that the first plants are a genetic fusion of the two. Ninety percent of plants have fungi called mycorrhizal—literally, "root fungus"—living in special co-evolved compartments inside their roots or in the soil, intertwined with roots, making food for each other

128

according to specialty and recycling wastes.

Meanwhile, worms and arthropods are the first animals to follow algae ashore. They probably find the protection of wet insides easier, given the armor they have already evolved. But walking about with the greater gravity pull of air is a problem for arthropods. Some grow stronger muscles; others solve the problem by evolving the nearly weightless bodies of insects, which even permit them to take to the air as the first flying animals.

The dry land of Earth's crust is beginning to reorganize dramatically. Eroding rock, mud, water, and the existing coat of bacteria are now joined by algae, fungi, worms, and arthropods in increasing numbers. All together this makes for a growing cover of soil. Biomass increases and land transforms into ecosystems.

It takes a long time for plants to develop proper roots and vascular systems—tubes or veins

---

**5 4 0  M I L L I O N  Y E A R S  A G O**

CAMBRIAN EXPLOSION

For years, we knew very little about the microcosmos. Bacteria and protoctists abundant for over two thousand million years remain soft-bodied. The great discoveries of Cambrian fossils were taken to imply an explosion of life from virtually nowhere. | A tremendous burst of animal evolution springs from protoctists and animal biomineralization. Life rapidly radiates. All of today's animal phyla (great groups) originate in the Cambrian period. Magnificent fossil traces (burrows) record our own distinguished, flexible ancestors—the worms.

---

2,000 MYA 1,000 MYA PRESENT

(above) This velvet worm is very similar to fossils found in the Burgess Shale. These animals are thought to be the link between two extensive and important phyla, the arthropods (which include all insects and seagoing crustaceans) and the annelids (which are segmented worms). Scientists use these fossils, widespread before the break up of Gondwanaland (precursor to our present Southern Hemisphere), to reconstruct the history of drifting continents.

(opposite) Opabinia, a fantasy-like predator of the Burgess Shale, measures three inches long. It has five eyes, gills all along its segmented body, and an efficient nozzle which vacuums prey for transfer to its mouth.

130

to carry water drawn through roots to all parts of their bodies—as well as taller, stiffer bodies that can evolve into ferns and trees. But as they do, they are always part of complex ecosystems of species in co-evolution: trees, smaller plants, lichens, fungi, and insects, covered in and inhabited by bacteria and protists.

Plants evolve ways of reproducing in the absence of water. Like animals, they develop ovaries, producing ova, or eggs, that are fertilized by pollen and develop into embryos within their protective bodies. Plants depend on insects for reliable transfers of pollen, while wind, birds, and other animals scatter their fertile seeds. The more we learn about evolution, the more we see how everything co-evolves interdependently.

About 440 MYA, just as early plants, fungi, and insects are flourishing on land and the seas are filled with life, the first great extinction on record occurs. Ancient climates and rates of

## 510 MILLION YEARS AGO
### THE BURGESS SHALE

The Burgess Shale, in the Burgess Pass of the Canadian Rockies, is an impressive fossil find dating a some 30 million years into the Cambrian period. Its impressions are especially precious because such preservation of soft-bodied marine animals is rare in the fossil record. This discovery provides a unique glimpse into the true range and diversity of early animal forms and their ecosystem.

biological activity—low ones pointing to extinctions—can be measured in ocean sediments of former biomass. Various proportions of different forms of carbon and oxygen incorporated by ancient life-forms are clues to such information and are recorded in biomass remains even after hundreds of millions of years. Such information tells us that this extinction is associated with another great ice age, when glaciers form and chill the land and seas.

More than half of all marine and land species are extinguished, apparently as a result of this climate change. It takes 25 million years for Earth to recover its biodiversity, much of it with new creatures. As we might expect, bacteria and protists find it easiest to survive, producing rich plankton soups even in the colder seas where they are happiest. This creates vast food supplies for both surviving and newly emerging larger species.

132

## Amphibians Ashore

It is during the recovery from this great extinction in the middle of the Paleozoic that animals other than arthropods emerge from the sea to venture ashore. For these fishlike creatures, the major biotechnology problems are the need to breathe air and the need to support their bodies in that thin medium. We can still see mudskipping

*Sand shark*

### 5 0 5  MILLION YEARS AGO
#### REVERBERATING CHORDS

Human beings belong to the phylum Chordata, all members of which have notochords (cartilage rods down the middle of the back) at some time during their life history. Chordata include all vertebrates (mammals, birds, amphibians, reptiles, and fish) along with several groups of lesser known marine animals. Details of the path in the evolution from notochords in sea-squirts to those in jawless fish to those in bony fish remain obscure. | Whatever the precise path, the backbone that enables us to stand speaks to the wonder of evolution: branches, truncations, mergers, as evolutionary continuity spans nearly 4,000 million years.

*Squid have nervous systems very similar to our own.*

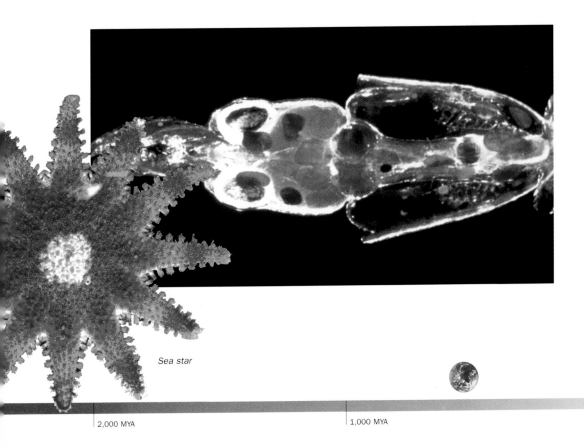

*Sea star*

134

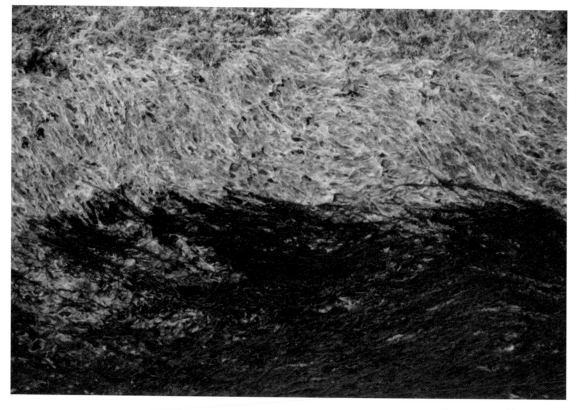

These green and brown algal protoctists are thready chains of cells full of plastids.

HAPPY LANDINGS *The Original Green Movement*

Green algae (ancestors to plants) living in sunlit shallows and tide pools start a "green" revolution. They invent ways to stay wet on the inside while drying up on the outside. Algae become plants by carrying the ocean inside themselves as they move ashore. | Why go to all that effort when they could just lounge at the sea's edge? Some algal colonies take to the land to avoid a disturbing increase in marine predators. Once ashore, the spirit of life drives them to expand across this huge new frontier. The insects, which are both predators and pals to algae, follow with dispatch.

fishes flapping along on shores, reminding us of their—and our—land pioneer ancestors. Species that continue a double lifestyle at sea and on land are called amphibious, literally meaning "double-lived."

Very likely this phase of evolution is driven by the same problem algae faced, that of being stranded when tides receded. Somewhere along the line, it must become easier to survive on land than to drag back to the sea after each stranding. Thus, the first amphibian animals hop and lumber along on ever more muscular fins while trading in their gill slits for air-breathing lungs. Eventually the fins transform into stubby legs, and a second pair is added to make them four-footed creatures, or tetrapods—the first in long lines

of amphibians, reptiles, birds, and mammals.

Our modern frogs and salamanders give us only a sample of the far greater variety of amphibians living in ancient times. But however long amphibians can live on land, they must all go back to water to lay their eggs. These hatch into swimming tadpoles that repeat the ancestral process of growing legs and coming ashore in every generation to this day.

Amphibians multiply and co-evolve with the early plants and fungi, but survival is challenged again. Less than 100 million years after the first mass extinction, another occurs around 365 MYA. As before, it seems related to climate change and reduces the number of species by half.

135

The amazing arthropods, which constitute over 80% of the species in the animal kingdom, are the first to follow algae inland. No doubt their hard-shelled, lobsterlike legs helped.

## THE COMING AND GOING OF CARBONIFEROUS FORESTS

During the thirty million years it takes to recover the loss, Earth again grows warmer. Hardy mosses and horsetails that survive to this day join with evolving tree ferns to form the first subtropical forests. We call this time—still within the Paleozoic—the Carboniferous period because, over millions of years, succeeding generations of these extensive forests load themselves with carbon extracted from the atmosphere.

As we now know, they are gradually pressed into the ground and fossilized as coal and oil—the fossil fuels we are digging back out of Earth with phenomenal speed, once again increasing the $CO_2$ of our atmosphere as we burn them.

Billions of years ago oxygen was the great danger. Today the danger is carbon dioxide. It took Earth's life-forms billions of years to pump the carbon dioxide from our atmosphere and transform it into oil deposits while filling the atmosphere with nitrogen and the oxygen we need to breathe. Now we reverse the process by pumping out and burning the oil, and by burning present forests with their own stored carbon. Earth's surface is heating up because the additional carbon dioxide prevents its normal loss of heat in a greenhouse effect.

Atmospheric $CO_2$ is rapidly approaching levels it apparently reached just before the last ice age. Perhaps Earth will cool her man-made fever with a new ice age, destroying most of what we have built and forcing us into retreat, like the ancient bubbler bacteria, to more hospitable environments. Or perhaps Earth will shift into the opposite mode to produce a hot age—also within her evolutionary repertoire.[5] In such an event, the polar caps would melt until only a few islands remained above sea level.

In the ancient carboniferous forests, endless associations and symbioses are worked out among life-forms in the continual recycling of water and nutrients—one creature's wastes becoming another creature's food. New species move opportunistically into new ecological niches, spreading their kind as fast as possible and competing against intruders. We can still watch this happen today when we clear ecosystems and fail to develop them—say, for example, a cleared field that goes unplanted. Later, the different exploiters of the ecosystem mature as co-evolving species, adapting to each other's needs over time and turning competition into cooperation, thereby forming intricately balanced ecosystems.

There is some evidence in coal fossils that tree ferns more than 300 MYA produced the same kinds of response to insect attacks that plants do today. Rather than letting insects feed on vital parts of their bodies, plants may respond to chemical signals from the insects to produce extra masses of cells, called galls, that house and feed the insects. Such galls occur on ancient tree ferns.[44] Perhaps this is another instance of competitive exploitation turned to cooperation.

Other such plant and insect symbioses known today include the protection against predators that some species of rain forest ants give to certain trees in turn for liquid food the trees produce exclusively for these ants. This is one good reason for being careful about which trees you touch in walking through a rain forest, for the ants are not obvious until the tree is touched and they spring to its rescue. Tracing the countless instances of co-evolved symbioses in nature is still a new venture for biologists. It reflects the growing trend away from seeing species as independent lineages and the recognition that ecosystems evolve as wholes.

As the carboniferous forests spread, amphibians gradually transform into pre-reptilian creatures. These emancipate themselves from the endless amphibian returns to the sea for reproduction by inventing the self-contained egg. This egg contains fluid as well as enough food to sustain the embryo until birth. The embryo with its food is encased in a tough protecting membrane or shell from which it can break free to hatch on land.

Tetrapods that lay eggs on land split into two lineages. One is the synapsids—meaning "with arch" and referring to the shape of the inner skull—that evolve into the dominant tetrapods by the middle of the Mezozoic era and ultimately into mammals and humans. The other is the reptiles, which in turn divide into three branches: the first, turtles and other bulky four-footed creatures that have not survived; the second, snakes and lizards; the third, archosauromorphs— "ruling lizard forms"—which include all our beloved dinosaurs as well as pterosaurs—"winged lizards" —crocodiles and birds. Synapsids are often confused with dinosaurs because early ones took on forms similar to those of reptiles.[44]

Reptiles evolve a grand variety of sizes, shapes, and habitats, including water, deserts, and forests. But before all this can happen, the 70 million years of warm and prolific Paleozoic carboniferous forests in which reptiles first evolve come to an end. The tectonic plates are assembling the continents that ride their backs once again, this time merging Laurasia and Gondwana into the single landmass of Pangaea, which means "All Gaia." Mountains are raised in the crunches and landscapes change; continental shelves break up and remaining shorelines around its edges reform into new marine habitats.

137

The part of Pangaea formed by Laurasia—later giving rise to North America, Europe, and Asia—lies northward from the south polar Gondwana, later giving rise to South America, Africa, Australia, and Antarctica. As they merge to form a supercontinent stretching from pole to pole on one side of Earth, ocean currents are interrupted and rerouted as the continental climate overall becomes colder and drier. Dramatic serial climate changes occur once again as new glaciations destroy the remaining subtropical forests and widespread deserts follow on each retreat of the ice sheets.

When Pangaea breaks up again later, plate tectonics eventually lead to the present arrangement of continents dominating the Northern Hemisphere. If we could watch tectonic plate activity like a movie from its beginning, we would see the plates riding the slowly swirling molten insides of Earth. It may be that animals

138

*Crucial to global metabolism, fungi transform waste and dead bodies into life-sustaining resources.*

CONCEPT NOTE

## WHAT ARE FUNGI?

Fungi have an unbounded love of life and death. Converting waste and corpses into resources, they are crucial to global metabolism. Inverting our habit of consumption, fungi digest their food before they eat it. They excrete enzymes onto organic materials and then absorb the soluble, predigested meals. | What we see in the wild is only the tip of the fungi; they spread gregariously underground. A famed Michigan fungus—one individual fungus with identical genes throughout —has been expanding for over 1,500 years. It spans 37 acres and weighs over 11 tons! | Unlike animals and plants which form embryos, fungi form propagules—dormant or reproductive environmentally-resistant spores. The propagules can be blown about for thousands of years before moisture

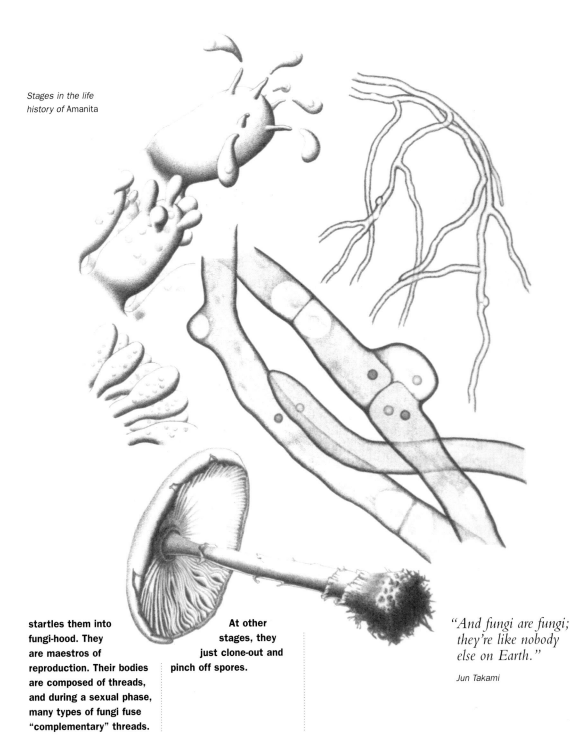

*Stages in the life history of* Amanita

139

startles them into fungi-hood. They are maestros of reproduction. Their bodies are composed of threads, and during a sexual phase, many types of fungi fuse "complementary" threads.

At other stages, they just clone-out and pinch off spores.

*"And fungi are fungi; they're like nobody else on Earth."*

Jun Takami

have by now affected the speed of this movement —in part by their responsibility for sediments heavy enough to push plate edges under each other more rapidly and in part by the greasing action of their concentrated calcium.

We can see why all this tectonic activity with its attendant climate extremes is harder on large plants and animals than on protists and bacteria, which to this day—despite the way the world looks to human eyes—maintain greater variety than the other kingdoms.

As Pangaea forms, hardy spore-making ferns and cone-bearing trees such as ginkos and cycads evolve in the cooler, drier habitats, assisted by legions of bacteria, protists, and fungi. They also establish symbioses with insects and with those amphibians and pre-reptilians able to survive the huge crustal crunching and the cooling climate.

140

## REIGN OF THE REPTILES

About a quarter billion years ago (250 MYA), near the end of the Paleozoic, a lizard we named *Coelurosauravus jaekeli* takes bravely to the air on weird wings resembling hang-gliding gear made of hollow tubes and light webbing. But before other reptiles can transform themselves into dinosaurs and avian creatures, Earth's flora and fauna succumb to the greatest of all extinctions around 245 MYA. Perhaps it is a kind of domino effect as key species in food chains fail to adjust to changes in climate. In any case, the Paleozoic era ends in tragic disaster as 95 percent of all species go extinct.

The time from 245 MYA to 65 MYA is designated as the Mezozoic—literally, "mid-animal" —era, following on the "old animal" Paleozoic and divided into Triassic, Jurassic, and Cretaceous periods. In the Triassic we see the recovery from

---

**4 5 5 MILLION YEARS AGO**

FUNGAL FUSION *Plant-It Earth*

The Plant and Fungi Kingdoms evolve on land so closely together in time that we are unclear which of the two came first. The fungal fusion hypothesis opts for neither, suggesting a joint venture.

It proposes that fungi and landward photosynthetic algal symbionts joined together to evolve the capability to survive dryness. Was it just a joint venture or did it go further? | Perhaps land plants

arose through both a somatic (body) and a permanent fusion of fungi with algae. This symbiogenesis is much like the "horizontal" gene transfer which bacteria practice. | Life on land poses far

different demands than did life in the sea. Lifeforms develop "hypersea" —the movement of sea water onto land, but inside organisms.

How many ecosystems are within this ecosystem?

"*Redwood trees, wasps and other organisms visible to the unaided eye are habitats for rich ecosystems of much smaller organisms that cover their surfaces, probe their interstices, take up residence in body cavities, and invade (sometimes peacefully, sometimes not) even the inner sanctum of their cells.*"

Mark McMenamin, *Hypersea*

141

this greatest of all extinctions, including the growing dominance of therapsids, derived from the more generalized synapsids, looking like sleek predatory dogs and large fat herbivores with beaks and tusks.[44] Among them live carniverous crocodilelike *archosaurs*, erect bipedal *dinosaurs*, and somewhat airborne *pterosaur* lizard-birds. In the oceans we see the emergence of everything from those spectacularly beautiful shell-building protists, such as the "Emilies," to a plethora of new, larger animals and spreading seaweed.

*Archosauromorphs* are evolving into a variety of creatures that eventually look like everything from large lizards—which they are not—to reptilian pigs—which they also are not. Among them are the dinosaurs that evolve into crocodiles—including seagoing crocodiles with flippers—and the pterosaurs that evolve into birds.

But in the late Triassic, only 37 million years after the great extinction bringing it about, yet another extinction occurs! This one takes half the recovered and new species with it around 208 MYA. The two extinctions together require 100 million years of recovery. Most synapsids and archosaurs succumb, yet the smallest of the first mammals giving birth to live young survive the new extinction to herald the Jurassic period. They live in a complex, if devastated, world of reptiles, amphibians, insects, trees, plants, fungi, and microbes. Tiny furry mini-mammals survive where larger ones die out. Rodent-size insect eaters are the first to nurture their babies on milk produced in their bodies.

It is after this extinction that diatoms make their appearance among oceangoing plankton, building their spectacular shells of opaline silica in huge quantities to assist, if not upstage, the Emilies in moving materials about and seeding clouds with their exhaust gases. When diatoms appear, the seas are supersaturated with silica, so

142

---

**CONCEPT NOTE**

WHAT ARE PLANTS?

Plants are more complex than animals and fungi: in addition to nuclei and mitochondria, plant cells contain plastids (organelles for tapping the sun's energy). Plants evolved from protoctist algae that had already incorporated cyanobacteria that became chloroplasts. | Significant challenges face the earliest plants as they confront the demands of surviving on dry land. Accustomed to an aquatic lifestyle, early plants lay on the surface, unable to support their weight against gravity. Watery conjugation is no longer a viable survival strategy. The most crucial innovation for the first true plant is the ability to develop a fertile egg into an embryo, a multicellular young plant, within moist, protective maternal tissue.

*Life history of an oak tree. Great oaks from little acorns grow. Sexual coupling of the female and male flower (left) results in the embryo.*

144

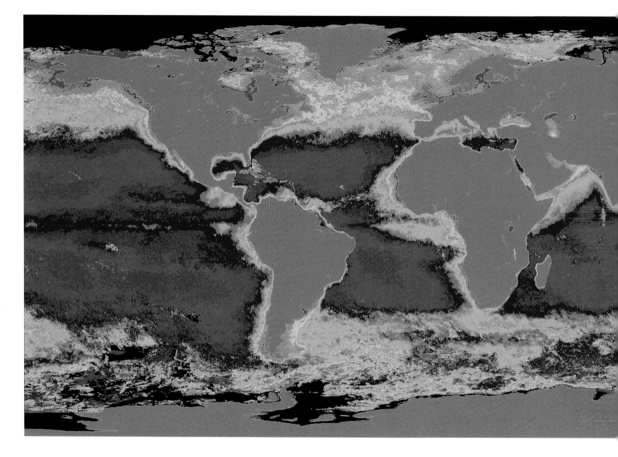

440 MILLION YEARS AGO

FORTUNE'S WHEEL

Global environmental change and continental glaciation induce a mass extinction which affects all marine animals. As the Ordovician period closes, over 50% of species worldwide decline and then vanish. The mass deaths open new niches for benthic (bottom-dwelling) and planktic (free-floating) marine life. New species and new groups of organisms which depend on these primary producers evolve. Researchers estimate biodiversity did not fully recover from this extinction for 25 million years. | Photoplankton (photosynthesizing protoctists) bloom in colder, resource-rich waters. They, in turn, enrich the food web. It is a period of plenty in the seas: plenty of room and plenty to eat.

Computer and satellite technologies make visible the ubiquity and power of ocean photoplankton. This composite, highlighting chlorophyll, shows ocean productivity. Note the blooms of life in colder waters.

they proliferate wildly. By now the population of diatoms adjusts precisely to incoming silica, helping to regulate the overall balance in ocean composition.

In the Jurassic and Cretaceous periods, dinosaurs and pterosaurs come to reign and spread all over the supercontinent so that they are found in all the parts breaking up. Dinosaurs grow up to forty meters long and some are the tallest beasts ever to live. Despite their fearsome appearance, most live entirely on plants, while fewer eat each other's eggs and smaller creatures.

Dinosaurs include huge-headed, horned ceratopsians such as *Triceratops*, which return dinosaurs to four-footed walking from their originally upright postures. The most giant vegetarians, or herbivores, *Brontosaurus* and *Brachiosaurus,* also walk on all fours as ponderously slow beasts, though they can still rise onto their hind legs to reach even higher than their long necks permit.

145

*Over 25,000 fungal species consort with photosynthetic companions, producing eminent varieties of lichens.*

146

## 4 3 5 MILLION YEARS AGO
### THE LICHEN CONSOLIDATION

Lichen land pioneers spread. Hardy and long-lived (some reach 9,000 years of age), these low-lying photosynthesizers are an arresting example of symbiosis. | Just as bacteria and protoctist mergers led to algae, lichens represent a merger of fungi with photosynthesizers (algae and/or cyanobacteria). An entirely new life form, lichens enjoy the algae's ability to use solar power to make food and the fungi's ability to store water and protect themselves from the elements. | Through rock weathering, lichens play a significant role in the geological cycle. Crustose lichens produce acids which chemically decompose rocky substrates; lichens manufacture a variety of acids depending upon the nature of the rock. Lichen also produce a glory of pigments.

147

(below) On the left is a healthy fungus; right, a healthy alga.
Their merger produced the British Soldier Lichen (below center,
also in photograph above left). Taking everything we know about
algae and fungi, we still never would have predicted the outcome
of their synergy.

**We still do not understand
how or why.**

Among them flit many varieties of tiny birdlike saurians and various intermediate forms. Theropods such as *Tyrannosaurus rex*—the "tyrant lizard king"—remain fully bipedal and have the huge sharp teeth of carnivores and carcass scavengers. Then there are the long-reigning ornithopodians with their tusks and duckbills, such as the *Iguanodon*, ancestor to birds.

The inseparable link between the smallest and largest creatures continues. Anaerobic bacteria by the billions feast away symbiotically in herbivore dinosaur guts, just as they do in aiding the digestion of cows and elephants, not to mention termites, today. Plants produce oxygen for animals, which reciprocate by producing the carbon dioxide plants need. Plants flourish to transform more and more solar energy into food for the demanding large beasts that fertilize them in turn. They also provide housing, as mammals curl up in tree root burrows or high in their branches. Animals, in turn, package seeds into fertilizer droppings, where they grow with adequate nourishment. Insects feed off plants and spread their pollen; early birds feed off insects and distribute plant seeds. Every creature, from microbe to *T. rex*, seems to have a part in the ever more complex ecosystems, as producers, consumers, rebalancers, and recyclers.

Scientists remain divided on the question of

whether dinosaurs were warm-blooded or endothermic, though much evidence based on form, growth rates, energy requirements, and behavior suggests that at least some of them may have been warm-blooded. The whole question has proved enormously complex, and is perhaps best summed up by the statement that evidence on dinosaurs "continues to suggest that they were something else: something neither mammalian, nor avian, nor reptilian."[45]

By the late Cretacious, large marine lizards called mosasaurs appear in the seas, along with "fish lizard" icthyosaurs, spiral-shelled ammonites, bivalve clams, and the lovely plankton foraminifera that add their share of mineral transport to the work of Emilies and diatoms.

Following the lineage from dinosaurs to birds via Iguanodon, we come to the well-known *Archeopteryx* and *Pterodactyl*—meaning, respectively, "ancient wings" and "feather fingers." Their fossils show us front leg bones evolving into wing bones with spans of up to twelve meters, jaws into beaks, and scales into feathers. This permits chasing insects through treetops, and moving far afield in the process.

Later birds evolve patterns of migration for climate comfort and steady food supplies. Birds become adept at reading nature's signals— orienting by stars and Earth's magnetic fields,

148

by the sounds of rivers, and the infrasounds of mountains. Their magnetic compasses are a replay and further development of ancient bacterial design.

In wings, the marvelous capability of tensegrity structures literally finds its highest expression, the strength of the lightweight muscles pulling tightly against compressible bone without breaking it. One of the fun as well as useful ways scientists have found to understand animal flight is to make giant flying models. From the time of Leonardo da Vinci, people tried long and hard to produce single-person airplanes that could be lifted from the ground by manpower alone. The problem was not solved until the late 1970s, with the California Gossamer Condor, made of the lightest and strongest material available—hollow aluminum tubes from which soft-drink cans are cut, covered in plastic kitchen wrap. Even so, the plane has to be pedaled like a bicycle to get it airborne. Flapping wings large enough to lift us with our arms is impossible.

Size can also be a problem for larger creatures that need huge amounts of food if climate changes reduce its availability. As we know, dinosaurs do not survive, while creatures whose fossils record miniaturization over time are more likely to survive mass extinction.

## ODE TO INSECTS AND FLOWERS

During the rise and reign of dinosaurs, another marvelous and intertwined success story is being written. Great forests that have grown up in the Cretacious period near the end of the Mezozoic era explode in color as flowers emerge in close cooperation with insects. Before we introduce the flowers, let us look at some developments in the insect world, including the remarkable beetles.

Beetles are four-winged insects that trade their forewings for beautiful armored shells. They are arguably the greatest success story in all evolution if we count by numbers. Earth has more beetle species than plant species. While birds and mammals together account for less than 1 percent of all known species, 60 percent are insects. One-third of those are beetles, ranging from microscopic sizes to half a foot in length, from inconspicuous dark colors to iridescence so dazzling they have been worn as jewelry.[46]

Up close, many insect species are models for science fiction, as were their more ancient Burgess Shale creature ancestors. They have shells of shining armor, huge multifaceted eyes, all manner of horns and antennae. Some can hurl each other about in battle and others look like long-legged mechanical Moonwalkers. Beetle

149

skills include feats of running at the equivalent speed of a horse doing 250 miles per hour; grasshoppers leap the equivalent of skyscraper heights on their own hind legs; ants haul unthinkably heavy loads; and cockroaches are the most touted survivors, with immune systems so like our own that we study them to understand ourselves better.

150

As the variety of insects increases, plants evolve gorgeous and overt sexual organs to attract them. We know these organs as flowers. Flowering plants are hermaphroditic, with both sexes in the same plant. Attracting insects, birds, and mammals to cross-pollinate them is thus a good evolutionary strategy promoting genetic variety. Flowers make themselves irresistible to insects by way of colors, perfumes, and delightful nectars, while insects learn to detect them with color vision extending beyond our human range, a delicate sense of smell, and a taste for nectar.

(above) The synergistic mycorrhiza root of an alfalfa plant, a symbiotic protuberance produced by fungus and alfalfa plant roots.

(opposite) A mycorrhizal root fungus from the Rhynie chert of Scotland, one of the world's most important fossil deposits

## 400 MILLION YEARS AGO
### INTIMATE ALLIANCES

Intimate ecological interactions occur among plants, fungi, and bacteria. Mycorrhizal fungi live within special root compartments co-created with plant partners—and they are symbionts with over 90% of living plants today. Fungi help make many valuable nutrients available to plants. The plants provide sugars to the fungi. | Symbiosis generates the high diversity and vast biomass of terrestrial life. All organisms consist primarily of water, and interact easily in fluid habitats. The evolution in land biota of the intimate association of networks of cells (through which fluids and solids are transported) are already well established in the Devonian period.

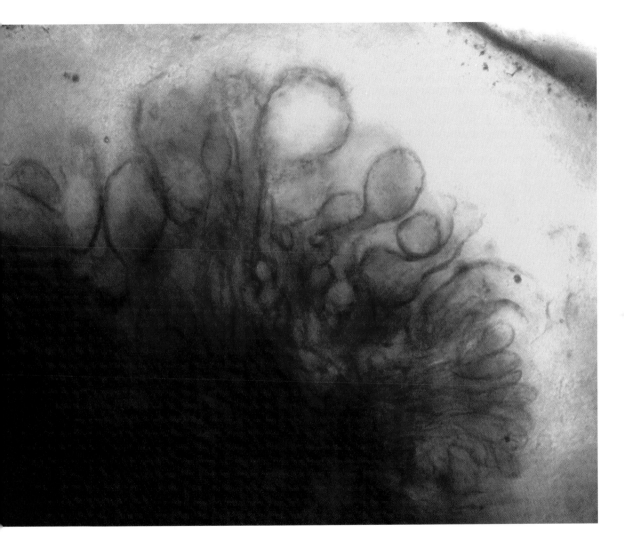

151

As they visit each flower, they gather and spread pollen. Flowers later go on to become fruits most appealing to birds and mammals, which package and spread their fertile seeds.

The relationship between insects and flowering plants contributes to the evolution of some complex communications systems in insects. Bees, for example, evolve a special language of dance to tell each other where flowers are located, each part of the specific dance indicating distance from the hive, direction, or other information about the flower supply.

Insects develop other intricate communications systems based on the perception of electromagnetic fields and chemical stimuli, including pheromones that they themselves emit to attract each other in their private mating dances. They also develop communications to warn of danger. For example, when a night-mating pair of Amazon spiders that glow in the act of love is

152

eaten by a snake, some cry of distress on the spiders' special frequency is perceived by others of their kind, which instantly blink off their own lights to avoid detection.

## AN ASTEROID BRINGS MAMMALS TO THE FORE

Suddenly, around 65 MYA, the Mezozoic ends, literally with a bang, as the entire world of reigning reptiles, mammals, and emerging birds in their luxuriant forests of trees and insects, flowers and fungi, is shattered beyond recognition. The disaster apparently is caused by a huge

*(opposite) Fish, evolving into amphibians, were the first vertebrates (the group of animals with backbones, to which human beings belong) to make it to land.*

---

| 3 9 5 | M I L L I O N   Y E A R S   A G O |
|---|---|

AMPHIBIANS *Lured to Land*

Air breathing, four-footed, ambling amphibians leave many marks by the late Devonian period. Their ancestors—the lobe-fin fishes—were most likely lured out of the oceans by a profusion of insects. |

Evolving to breathe in air was not the only challenge faced by lobe-finned fish in their move to land. They also had to support their weight against gravity. The bony skeletons of amphibian precursors

(who lobbed about on already muscular fins) give clear clues to the transition some animals made from dragging in drying mud pools to true walking movements. | Amphibians do not make a complete

land transition: they must return home to lay eggs, where their tadpole progeny keep one evolutionary foot in the water.

153

154

RYTHMS OF LIFE AND LOSS

The Devonian period closes with another decline and mass extinction. Again, 50% of species vanish worldwide, the major losses taken this time by ocean life. Basic biota body blueprints remain conservative. | Note the "bottom-heavy" trend in evolution: new lineages generate remarkable diversity when they first appear, but settle to a limited number of body plans in what paleontologist Stephen J. Gould calls "early experimentation and later standardization." | It takes 30 million years for biodiversity to recover fully from this mass extinction.

asteroid hurtling from the skies to splat down violently into the area we now know as the Yucatan Peninsula of Mexico. First incinerating crustal life-forms with its debris, the asteroid then freezes those remaining by raising a dust cloud that blocks the Sun.

The dinosaurs that have dominated landscapes around the world succumb, along with early large mammals such as therapsids, and the giant airborne pterydactyls and acheopteryx. Even sea animals are hard hit as most of their species disappear. Yet nature once again rises from the ashes to bring a positive side to disaster: the absence of the largest creatures opens pathways of evolution opportunity for smaller creatures.

The Mezozoic has ended and the Cenozoic era that brings us to the present has begun. Surviving mammals that have been active only by night, including small primates, come out of their trees, caves, and burrows to live by daylight. Their stereoscopic vision adjusts to the brightness as they move about on agile limbs, plucking their food with dexterous digits, carrying young upon their backs, and weaving their social systems into communities. Plant food is more plentiful than ever with the great grazing reptiles gone, though pollenating plants disappear for a while. Edible roots and shoots abound in leafy fern-rich forests while flowers and fruits make their comeback.

Plants are anything but helpless under the onslaught of new herbivores. If they need protection, they enlist insects, as we saw earlier, evolve their own poisons, or grow thorns. Some are less obvious, evolving leaf acids directly triggered by mammalian chompings, turning their initially sweet leaves bitter in minutes. This insures that animals do not overgraze any particular plant's leaves to do serious damage.[48] Birds evolve beaks suited precisely to their preferred food—short and strong to crack seeds, long and delicate to

155

(opposite) Exemplary of the way new life fills evolutionary niches, life grows in a tide pool. Climate and changes in the nature and distribution of habitats appear to drive most of the recent mass extinctions and extensive speciations which follow. Species capable of filling newly emptied niches do so rapidly.

pick insects out of their holes, and even with storage sacks to hold large fish. Four-legged animals in turn get wily about equipment for finding food, grow fur appropriate to season and habitat, and protect their young in ingenious ways.

Mammals branch into two basic types—marsupials, such as the kangaroo and the opossum, which give birth to very undeveloped young, keeping them in pouches on the outside of the mother's body for further development, and those that develop the embryo fully within the mother's body.[49] But not all mammals become fully mammalian. As if to confuse scientists, the platypus is a furry, warm-blooded animal, but it lays eggs and has a beak and webbed feet, as if it could not quite decide whether to be a bird or a mammal.

While baby reptiles hatch with agility and competence to go about their lives on their own, baby birds and mammals are born helpless and in need of shelter and of maternal, and sometimes paternal, care. Growing up, they show increasingly complex behaviors over time, including elaborate rituals and vocalizations as well as skilled hunting and nest building.

The flexible bodies of animals such as cats and primates work well for snuggling kinds of baby care and for tumbling in play, while hoofed animals such as goats and gazelles must rise to their feet earlier, run swiftly, and frolic differently. Each animal species is tailored to a lifestyle and an ecosystem because they have co-evolved.

Territoriality rituals evolve among birds, fish, and mammals, the usually male intraspecies rivals often flashing colorful body parts as signals in elaborate dances. They avoid combat to death and insure adequate space and resources for raising their families by engaging in these lively border disputes and negotiations. Such ritual behaviors

**3 6 0 MILLION YEARS AGO**

CARBONIFEROUS PERIOD *The Coal Forests*

Continental movement folds the lands. Extensive forests of mosses, horsetails, and tree ferns rise in massive basins during the sultry, swampy Carboniferous period. These plants practice "giant-ism." Some of their descendants, today's club mosses, will follow an alternate evolutionary strategy: when things get tough, get smaller. | Dead vegetation does not completely decay in these swamps. The dead organic matter accumulates in huge "carbon sinks." The burning of fossil fuel (coal, oil, and gas) during the 19th and 20th centuries has already consumed a substantial fraction of the fossil fuel laid down during the 70 million years of the Carboniferous period. This combustion has significantly raised the carbon dioxide content of the atmosphere, risking greenhouse warming of Earth.

157

The discovery in the Antarctic
of coal, which forms only
in warm, wet subtropical
settings, initially startled and
confused scientists. Ultimately,
however, it was one of many
observations that may be best
explained by a most important
earth system theory: continental drift and plate tectonics.

are situationally triggered and do not require learning—we say they come "wired in" as "fixed action patterns" inherited genetically. Humans have lost this animal heritage of sharing space through ritual combat without killing each other, having no inborn limits on our warfare with each other! The price of freedom in human behavior is sometimes high, and we must design ethical systems to keep ourselves voluntarily in check.

Birds and mammals evolve many of their complex social behaviors in connection with caring for their young: feeding, cleaning, vocalizing to find each other, protecting from harm, and teaching self-feeding. Mammals also evolve play, beginning with that between mother and baby, and among siblings. Much of it appears to be practice for later hunting and ritual combat. Mammals also evolve feelings of affection, insult, and other emotions.

Sleeping and dreaming are further special

158

## 340 MILLION YEARS AGO
### ENVELOPING EGGS

Amphibians transform into reptiles with a grand innovation: internal fertilization, which results in the "closed" egg. A new class of vertebrates spins off, generating the Great Age of the Reptiles. The reptiles' new reproductive strategy allows them to move inland to drier territories, where they rapidly expand the vertebrate presence on land. | Egg architecture evolves. A biomineralized shell prevents fluids from evaporating and protects the growing embryo. Separate compartments for pantry and for waste are easily accessed by the embryo.

(right) This herbivorous reptile, just under two meters in length, represents the transition from amphibian to reptile.

(opposite) Reptile relations—lizards, snakes, crocodiles, pterosaurs, and birds—later take advantage of the egg innovation. Communication systems emerge that become critical to chicks' survival. Unhatched chicks hear one another chirping and clicking, which somehow enables the baby birds to hatch at the same time. While still in the incubating egg they recognize parental alarm calls in such a way that they hush up until they receive a parental all-clear.

159

behaviors mammals evolve, and scientists are still not sure just what roles our nightly hibernation plays in its evolution. Does it keep early mammals quiet during the dinosaur-dangerous days? Does it prolong our lives? Hibernating bats live about five times as long as their same-size hyperactive shrew cousins, both having about the same number of heartbeats per lifetime.

With tall trees, dinosaurs, and large mammals such as mastodons and whales, Earth's creatures apparently reach their size limits. Trees that grow too tall cannot stand upright in storms or pump water to their highest leaves. Mammals the size of whales and elephants seem to be limited by their size for heating and cooling, getting and processing food, and maintaining their metabolism. When we multiply an animal's size by ten in each direction, its weight increases to a thousand times more and so it needs a thousand times more food. Perhaps this is why so few large animals evolved in comparison with smaller species.

Interestingly, there seem to be far fewer limits on the diversity of creature design—something truly astonishing considering the tiny handful of elements from which all are made. Animals evolve a dazzling variety of form and function along with sensory equipment for seeing, smelling, touching, tasting, feeling, radar tracking, and otherwise perceiving their physical, chemical,

and energetic worlds. They evolve endless and marvelous ways of swimming, slithering, running, climbing, and flying; of hunting, making homes, breeding, rearing young, playing, communicating, and ever learning new ways of doing things.

## MYSTERIOUS REVERSAL: A MAMMAL RETURNS TO THE SEA

Pangaea has been breaking up since the late Mezozoic, again separating Laurasia from Gondwana, forming an ancient ocean passage and then ocean called Tethys. About 50 MYA, Tethys is being squeezed into a warm and shallow sea as the subcontinent of India approaches Asia, pushing the seafloor upward. Apparently rich with plankton and other foods, it teems with life and attracts certain mammals to aquatic lifestyles. Recent evidence strongly suggests that wolflike creatures live on seashores that end up as what is now India and Pakistan. Are they scavengers venturing into the sea in search of food? What makes them utterly abandon land life to evolve flippers in place of their legs and transform ultimately into whales?

The evolution of cetaceans—our collective name for whales and dolphins—is not yet entirely clear. It appears that an evolutionary sequence of some 50 million years runs from these hyena or wolflike creatures to a kind of reverse evolution into hairy crocodilelike reptilians, such as *Rhodocetus*, and seal-like creatures that look like amphibians, such as *Pakicetus* and *Ambulocetus*. Then it seems they go on to become fully aquatic carnivorous predators that still have hind legs, such as *Dorudon atrox* with its ferocious teeth, and ultimately full-fledged whales and dolphins. Besides giving up fur and legs, and besides streamlining their bodies for swimming, many other physiological adjustments must be made, including the transformation of nostrils into blowholes and the adjustments of blood systems for deep diving. It is clearly not a simple transition and the fossil record leaves many questions.[50]

However it happens, whales and dolphins, sea lions, and other pinnipeds—literally, "finfeet"—flourish in the seas to this day, except for serious threats to their livelihood and survival by human hunters and polluters. Cetaceans, with their impressive brains, even larger than our own, thrive in great ranges of temperature and pressure, roaming our watery planet over extreme distances, some communicating over entire ocean basins. We are only beginning to study their songs and social lives.[51, 52, 53]

Today's whales are divided into baleen whales, such as humpbacks, which feed on plankton, straining it by the ton through the brushlike baleen growing from their upper jaws, and toothed whales, such as sperm whales, narwhals, and dolphins, which feed on fish and squid.

As whales and dolphins evolve in the Cenozoic, continents continue to separate, creating new ocean currents and seas to lure them into roaming ever more broadly. Eventually, three-fourths of the world is theirs. Africa moves ever farther from South America, and Greenland separates itself out of the Northern Atlantic region as Europe and North America move apart. India has "floated" from Antarctica to Asia, where it finally crunches into that huge landmass, raising the Himalayas as it does. The separating continents have stranded species to continue on different paths. Australia in particular develops its unique kangaroos and koalas, while alligators and crocodiles distinguish themselves in North America and Africa respectively.

Inland seas shrink. The Mediterranean Sea alternately dries out into a desert and floods again at the Straits of Gibraltar. Lake Titicaca's waters are raised high into the newly formed Andes as the Pacific plates crunch to complement the Atlantic spread. North America is alternately flooded and dried out. If all this could be seen as a short film, Earth would look like a giant living cell reorganizing its component parts.

By now our planet looks much as we see it

161

today in photos—shimmering blue oceans, rich green cover over much of the landmasses, weather patterns of swirling white clouds. With infrared we would see huge blooms of plankton riding the seas. Life is everywhere, abundant and apparent. Earth itself looks like a great gossamer cell in its atmospheric membrane.

## LIFE AS A NEGOTIATING PROCESS: HOLONS IN HOLARCHY

Looking more closely, we see how Earth's creatures are interwoven within and among species, within and among ecosystems, as they have been from the time archean bacteria evolved. Forests and beasts alike branch into diverse new forms and functions. Saber-toothed tigers, woolly mammoths, and huge bucks sporting enormous antler racks stalk among smaller creatures with which they co-evolve as we move toward the present,

where we can see the patterns of creature interaction within and among species most clearly.

Migrating geese, for example, have marvelous systems for rotating leadership and supporting each other's efforts and well-being as they fly in formation. They take exquisite advantage of wind currents, putting the least possible strain on themselves in their annual moves from one ecosystem to another. Fish and hoofed land-grazers school and herd in similar ways, many with interwoven social lives. Elephants have elaborate social rituals and behavior, also serving other species by trampling paths though dense vegetation, permitting smaller animals access to watering holes. Without them, whole ecosystems collapse, showing us how important each role is.

Birds of all sorts pick annoying insects from

*The giant ferns did not make it through the climate change—smaller close relatives did. Opposite, spring fern in serpent disguise.*

---

### 2 8 0 MILLION YEARS AGO
#### SOWING AND REAPING

Profound changes in climate and glaciation, and finally widespread desertification, are linked to the tectonic assembly of the supercontinent Pangaea. Conifers, ginkgos, and cycads— naked-seed, "cone-bearing" trees—and dry-tolerant spore ferns replace the dying lush Carboniferous forests. Wind carries pollen from tree to tree, a common pollination method before some animals begin to provide the service. | Cycads grow well in soils extremely low in nitrogen. Cyanobacteria, living in roots where they induce a special layer of tissue, convert nitrogen from the air to a form usable by the cycad.

163

mammalian bodies, as small fish do for large ones. Take the birds away and the mammals become infested to the point of illness. Insects have complex relationships with flowers, such that one cannot live without the other. Wherever we look in nature we can see endless symbiotic arrangements as species feed, fertilize, and clean up after each other.

Co-evolution is shown especially clearly in close interspecies relationships, such as the important predator-prey arrangements in which the benefits are not all on the predator's side. Prey species stay deft and healthy avoiding their predators, as well as by sacrificing their weakest members. Ancient predator-prey relations can sometimes be deduced when a species runs significantly faster, for example, than its present situation warrants—suggesting that either its fast-running predator or prey has gone extinct, leaving it with a useless ability.

The study of ecosystems reveals that living entities and systems evolve within each other. The protist *Mixotricha paradoxa*, for example, lives in the hindguts of termites, which in turn live in towering termite communities scattered across an ecosystem. Each termite contains a trillion bacteria inside 10 million protists. Thousands of such termites live in hundreds of tower communities within a single ecosystem.

Arthur Koestler gave us a conceptual model for understanding living entities as embedded within each other like Chinese boxes or Russian dolls. He called it "holons in holarchy."[54] Other holistic modelers have adopted it for its usefulness in describing the relationships within and among living systems.[55, 56] Everything in nature can be seen as belonging to such arrangements—molecules within cells within cell communities within ecosystems, or molecules within cells within organs within organisms within families within

164

## 2 6 5 MILLION YEARS AGO
### RADIANT REPTILES

Reptiles evolve many new species. Reptile fossils are abundant: aquatic reptiles, "stem reptiles," early ancestors of snakes and lizards, ancestors of turtles, and archosaurs are the first in line of the great ruling reptiles to come. The first mammal-like reptiles also appear. | Most paleontological excursions through time focus on large land animals. Ocean water, however, which is Earth's largest habitat, also bustles with life. As tectonic plates move and collide, altered landmasses, with their nutrient-enriched coastal habitats, appear and disappear.

*(opposite, above) Mammal-like reptiles are part of the bushy speciation of reptiles.*

*(opposite, below) Variations on the theme of mollusk: nudibranch (left) and "ole blue eyes," a scallop (right)*

165

*(left)* Coelurosauravus *in flight*

*(below) Ponder this amphibian's future.*

166

**250** MILLION YEARS AGO

## LEAPING LIZARDS

New forms of life launch into the air, occupying previously untapped habitats. Although birds, bats, and insects are aeronautical experts, no Earth organism is known to spend its entire life in the air.

*Coelurosauravus jaekeli,* one of the first known reptiles to take to the air, has a "totally bizarre" wing design. With membranes connected to hollow-rod skin structures, the fossil lizard's wings more resemble hang-gliding gear than the transformed forearms with which birds and bats keep themselves aloft. Based on a 1997 find, paleontologists suspect that the hang-gliding habit allowed the lizard to sail following a running or falling start.

communities within species within ecosystems. Rivers are entities within watersheds within larger bioregions, including the atmosphere. Thus we become aware of the rich interdependence among levels of natural organization.

This kind of understanding shows clearly that embedded systems, or holons, must contribute to the health of the holarchies embedding them if they are to remain healthy. They must continually balance their independence with interdependence, their autonomy with their holonomy, or membership in the holarchy. An easy way to see this is to think of marriage or family as two-level holarchies. Each person must negotiate his or her autonomy not only with the needs of the other individuals, but also with the needs of couple or family as holons in their own right. And this negotiating process is dynamic and continual, not only in our social relationships but everywhere in nature.

This is a very important conceptual breakthrough in the understanding of living systems in evolution. In your body, for example, each cell must look after its own well-being. Yet it must also make negotiated compromises with the organ and body that embed it, because its health depends on the health of the entire holarchy that is your body. In short, simultaneous self-interest at every level of a living holarchy leads to cooperation by way of dynamic and continual negotiation among all levels.

Charles Darwin, to whom we owe our pioneering concepts of evolution, proposed that competition among individuals was its driving force. Later evolutionists pointed to cooperation within species and thus held species themselves to be in competition with each other. Richard Dawkins opposed both views by holding the "selfish gene" to be the competitive element driving evolution. In a holistic view, we can see that it is really not a question of choosing one of these theories over the others, but seeing that each is about the self-interest of one holarchic level. This simultaneous self-interest creates a dynamic tension that can only be resolved by cooperation if the living ecosystem is to remain healthy.[57]

In other words, all three theories together make better sense of evolution than any one of them alone. Think about the termite example above in these terms. Could the bacteria or the protists in their guts, or the bacteria inside the protists, survive if any of them met only their self-interest without compromises toward the health of the holarchy?

We know the mitochondria cooperating in our cells long ago worked out a mutually consistent way of life with other cell parts. We know that we have a mutually consistent arrangement

167

with them, as we provide their fuel and safety while they make our energy. We see how all species, including our own, must work out their mutual consistency with one another as co-evolving parts of their ecosystems.

The dynamic balancing of autonomy and holonomy is a very important issue in our own current evolution as a species. We are truly challenged to manage our human affairs within Earth's holarchy as well as the ecosystems embedding us. Will humans prove to be as creative as ancient bacteria in the face of the pollution and overpopulation problems we have made?

Tracing the story of evolution, we see ever more complex ecosystems weaving themselves through such negotiations as plant and animal species evolve particular bodies and ways of life that complement each other and the ever-hardy bacteria, protists, and fungi. Each ecosystem runs on sunlight, conserves and spreads water, builds

up fertile soils, and helps maintain overall temperature. As Lovelock observed, Earth functions as a physiological entity of geobiological activity. Creatures create habitats as habitats in turn nudge creatures to adapt to their changes.

It is still new for us to understand evolution as co-evolution—to see it holistically from a systems perspective. Our formulation *of* life on Earth is giving way to its reformulation as the life of Earth. And the more we see the life of Earth holistically, the more we recognize its self-organizing creativity. As Eshel Ben-Jacob proposes: "The emergence of the new picture involves a shift . . . in which creativity is well within the realm of Natural sciences."[24]

*(opposite) Ocean drilling, ice core drilling, and computer modeling technologies enable us to move confidently onto the terrain of paleoenvironments. Global climate alteration figures in virtually all mass extinctions. As sea levels change, both cooling and warming modify land and ocean habitats.*

168

---

**2 4 5  MILLION YEARS AGO**

PERMO-TRIASSIC EXTINCTION

The Permian period ends with by far the greatest of the five mass extinctions occurring between 440 MYA and the 20th century. Over 95% of species and 50% of families disappear, roughly twice as many as in any of the other four: the Ordovician (438 MYA), Devonian (367 MYA), Triassic-Jurassic (208 MYA) and Cretaceous-Tertiary (65 MYA). Profound waves of life expansion and species diversification follow each mass extinction. | The first four mass extinctions involve dramatic declines, for which scientists have found no obvious cause, whereas the Cretaceous-Tertiary extinction likely resulted from a cataclysmic asteroid impact. | Full biodiversity recovery time following past mass extinctions ranges from 10 to 100 million years. If irreversible species loss,

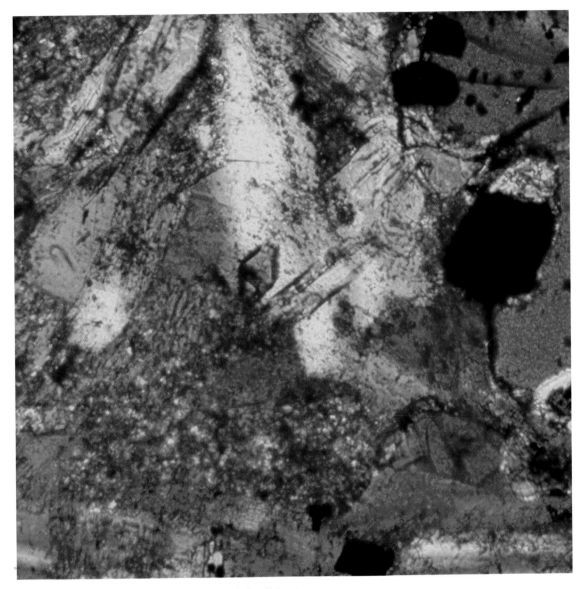

169

precipitated by the impact of human population sizes, continues at its current rate through the 21st century, Earth risks losing as many as one-third of all now living species worldwide by the year 2100.

Will we allow this insult to the non-human world?

## THE GREAT BRAIN EXPERIMENT

We are rapidly approaching the evolutionary time for our own appearance on the scene. To prepare for our emergence, let us pause to look at the evolution of our most distinguishing feature, our large, self-reflecting brain.

Because of the remarkable capacities of their brains, humans play an explosively sudden and utterly unique role in the drama of evolution—an experimental role, a part we write as we go, its direction and outcome yet unknown.

How do such brains come to be? We see that in the course of evolution, the brains of animals evolve in complexity to meet the demands of ever-increasing complexities with which they must cope. Brains evolve a step at a time from the first neuron to the simple neural rings that coordinate the swimming motions of jellyfish,

170

| 2 3 0 | M I L L I O N   Y E A R S   A G O |
|---|---|

EMILIANIA HUXLEYI *Belle of the Ball*

"Emily" is a planktic pro-toctist; this photosynthe-sizing alga spends time freely floating in the upper layers of the ocean, gath-ering solar energy. Though only a half a thousandth of an inch in diameter,

*Emiliania huxleyi* plays an expanding role in Earth's climate through both coccolith formation and gas emission. | A major geological force, proliferat-ing populations ("blooms") of this alga extract carbon

dioxide from the atmos-phere to form calcium carbonate shells that ultimately settle to carpet the sea floor, covering areas larger than all the continents put together. | In "bloom," Emily's gas

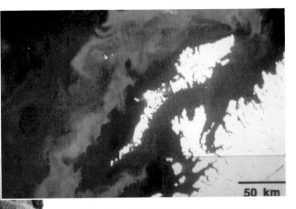

50 km

(above) The microbial 50-kilometer-wide bloom extends 200 kilometers along the coast of Scotland. When satellites first picked up these images, boats immediately went out to explore, but the shell coccoliths of Emiliania's shelly body were invisible to the unaided eye. Under a laboratory microscope, Emily and cohorts appeared aplenty.

(left) This SEM (scanning-electron micrograph) shows off Emily's almost baroque "coccoliths" (buttons). This environmental activist designs and bio-manufactures one plate every two hours, sending each to its proper place on the outside of the cell.

171

emissions are equally potent. As their sulfur-containing gas wafts into the atmosphere, solar radiation transforms it to sulfuric acid. The droplets of acid serve as nucleation sites for water condensation and the formation of ocean cloud cover.

and on via the first tiny brains of primitive worms to the superbly complex "networks" inside our own skulls.

We know that as brains become more complex, they also grow progressively more centralized and holarchical in organization and structure, just as individual cells did when eukaryotes evolved. This increases their levels of control and the range of voluntary behavior an animal can initiate. With exquisitely detailed brain development, for example, some ritualistic "innate" behaviors seem to loosen up, giving animals more flexibility and choice in responding to their environments and to their own inner urges. We can see this evolving flexibility in the increase of voluntary behavior, and in the ability to plan behavior based on stored information about past events. Eventually we call it conscious behavior. Human conscious behavior, of course, includes the development of symbolic thought and

language, which enormously increases complexity in social organization, both in geographical space and in time through generations.

Brains are often compared to computers, but no brain has ever been assembled from parts as we assemble computers. At every stage of its evolution, the brain is a fully integrated system derived from the single fertile cell beginning its organism's life. Evolving brains have the remarkable capacity to modify themselves as a result of experience and environmental demands. We know in our own case how memory and learning give us new abilities that are actually encoded into the brain after it is formed. After all, our brains evolved in their present form long before they were called on to write symphonies, build spaceships, or understand themselves. At birth, they are beautifully and intricately organized, but in so generalized a form that they constitute a massive soft potential waiting to be molded and

172

## 225 MILLION YEARS AGO

NIGHT JOURNEYS *First Mammals*

Small and nocturnal, the first mammals jump, climb, swing, and swim through the dinosaur world. Obliged to inhabit small niches in a world of giants, mammals will discover that their diminutive size opens a proverbial window of opportunity. | In the first waves of mammal diversification, some rodent-sized insect-eaters evolve lactation, enabling mothers to spend more time in the nest keeping their young both fed and warm. Some mammal species evolve even smaller bodies.

These insect-eating mammals, among the oldest known,
grew to about 4 inches (10 cm) in length.

inscribed by specific experience and learning.

The brain "knows" exactly how to organize, or "wire," itself. With an uncanny sense of direction, each growing neuron finds precisely the right path, say from the retina of the right eye to its particular point on the visual cortex, right next to an incoming neuron from the left eye. Embryonic brains actually have more neurons than fully developed organisms, eliminating many cells as more and more specialized connections are made. If some are not used within a critical period of time after birth, they will be eliminated, as shown, for example, by repeated studies in which kittens deprived of vision in early life cannot learn to see later.

A single neuron in our brains typically has some ten thousand physical connections to other neurons, and can apparently make or break them as needed. Even the identical clone offspring of a simple daphnid flea that can reproduce parthenogenetically—from a single mother—show different numbers and configurations of connections among their brain neurons, though each has precisely the same number of neurons. Perhaps this accounts for why brain neurons do not reproduce like other body cells, but turn over their molecules for new ones at a very high rate. That way they are always in good condition and never lose the intricate information patterns they create.

Human brains have the highest brain-to-body-weight ratio of all brains. To permit us to be born through the opening in our mothers' pelvic girdle, our brains and skulls are necessarily incomplete and malleable at birth, achieving a significant part of their growth after birth. This is one of the significant modifications made as we descend from apelike ancestors to engage in a whole new big-brain experiment.

174

## 208 MILLION YEARS AGO

SILICON SYMMETRY *Triassic-Jurassic Mass Extinction*

Just 37 million years after the Permo-Triassic mass extinction, the Triassic period comes to an end with another mass extinction. It will require 100 million years for biodiversity levels to recover from the combination of these two neighboring devastation of life. | New species evolve. Diatoms with magnificent silica microshells appear and spread quickly throughout the seas. These micro mineral-magicians extract and cycle silica and other elements from the oceans. | Before these creatures evolved, the oceans were supersaturated with soluble silica. As each silica-forming group arose, silica concentrations decreased. Major accumulations of biogenic opal begin to cover worldwide ocean floor.

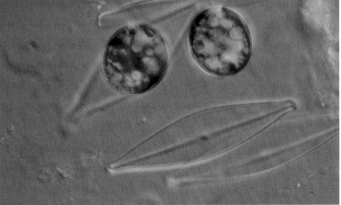

*(above) Each spine in this colonial diatom comes from an individual diatom cell. At one end of the colony, one cell operates as a steady grip, while all of the other cells float in the current and collect nutrients to share.*

*(right) These diatoms are mating. The gametes (reproductive cells), having left their silicious tests (the flying-saucer-shaped shells in the backgound), are about to fuse.*

## OUT OF THE TREES AND INTO THE TECHNOLOGICAL AGE

Quite naturally, the most fascinating story in all evolution to us humans is that of our own very recent appearance upon the scene. Some 13 MYA, orangutans branch off from the lineage of our own ape ancestors. Eight MYA, the lineages that are to become gorillas and humans branch apart, yet that is recent enough to leave our expressed DNA over 99 percent the same as that of gorillas. Around five MYA another branching led to today's humans and chimpanzees, our nearest relatives as evident from the nearly 99.9 percent of expressed DNA we have in common.

How shall we interpret this? Does it mean that our differences are only slight, despite appearances to the contrary? Or does it mean that DNA is less the determinant of structure and function than we have thought? Certainly

176

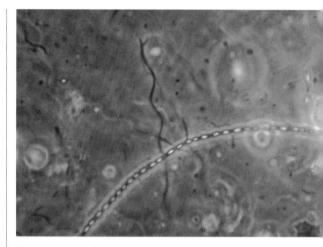

*This spore-forming filamentous bacterium (Arthromitus) lives in the anoxic world within termites and cochroaches. Its chain of cells attaches to the gut wall. Other kinds of unattached gut microbes swim upstream to avoid being defecated.*

| 1 9 0 | **M I L L I O N   Y E A R S   A G O** |
|---|---|

DINOS *Dining Out and Dining In*

**Modern American movies depict dinosaurs as dauntingly ferocious carnivores. In fact, most mega-massive dinos, regardless of their fearsome horns, spikes, and claws, eat plants. They feed on tough** cycad leaves, twigs, seeds, and fruits of trees common in shrublands and woodlands. Coprolites (fossilized plant matter in dino dung) show that while some herbivores mingled tastes, others **were fussy eaters. | Thousands of millions of cellulose-fermenting bacteria enable herbivorous dinos to digest the daily tonnage of cellulose (a process similar to that in 20th-century cows,** elephants, and termites). In exchange for their food-processing services, the microbes receive a large tract of habitat.

177

The paleontologists' menu features high-tech explorations into dino dietary preferences. Research on fossil bones and teeth adds considerable information about the dino taste in food.

"...The dimensions of bacterial biotransfer processes have considerably increased with the development of animals and plants; they, in turn, serve as energy sources and more fractal space for bacteria."

Wolfgang Krumbein

175 **MILLION YEARS AGO**

## ALL CREATURES GREAT AND SMALL

Huge dinosaurs rove mid-Jurassic Earth. Bigger is not necessarily better: larger life-forms require more space and food, and have fewer offspring and fewer survival options in times of change.

Microscopic "water bears," which survive today, appeared 200 million years before the huge dinosaurs. The tiny Californian water bear loves to cling to moss and lichens with its tiny claws.

The dinosaur and the water bear are archetypes of animal mega-micro waverings. Getting smaller is not uncommon in evolution. The fossil record suggests that beings which "miniaturized" are those most likely to survive mass extinction crises.

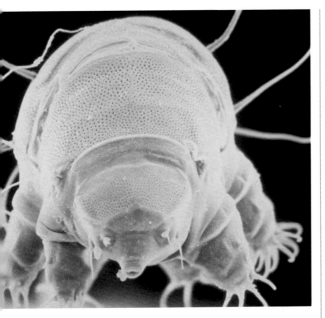

evolution's dramatically different rates of change are evident in us as speedsters. If polyps and cockroaches are like bicycles still functional in a jet age, then surely we are the jets themselves!

Around three MYA, a new ice age begins, with rhythmic glaciations once again alternately warming and cooling the planet. They also alter coastlines, sometimes forming land bridges—as between Siberia and North America —sometimes taking them away again. Mammals wander far and wide across Earth, but as far as we know, the largest primates continue to evolve and remain in Africa.

The first hominids, such as *Ramopithecus punjabicus* and *Australopithecus*, clearly still look like great apes. According to our best current knowledge, hominids move out of Africa around one million years ago, but do not really take on distinctively human facial features or lose their woolly coats until between 500,000 and 100,000

179

(opposite) Sauropods, reaching 120 feet in length, are representative of the giant dinosaurs of the Jurassic period.

(above) Tardigrades ("slow-step") or water bears, tolerate extremes. Water bears can survive almost total desiccation: rolling up into their tuns (mini-wine-cask shapes), they can hold out as long as 100 years awaiting water. They tolerate temperature ranges from 151 degrees C to -270 degrees C (almost absolute zero). They tolerate X-radiation: the lethal dose of X-rays for human beings is about 500 roentgens; for water bears, 570,000, making the bears of interest to scientists working on future space travel. They are tough, and many species can reproduce parthenogenetically (females hatching females on their own).

years ago. By this time their skills extend beyond foraging vegetable foods and eggs or scavenging animal remains left by other predators. They begin making tools and weapons, stalking larger prey in teams using tools made with deft opposing-thumb hands. On occasion they may bring down an animal as large as a mammoth, though most of their meat probably comes from much smaller netted mammals, birds, and fish.

As we become fully human, our bodies become more upright and our faces change significantly. Earlier in this story, we encountered the evolutionary concept of neoteny, when a juvenile form of polyp branched off into a new lineage as free-swimming creatures. We hinted then that human neoteny is apparent in noting the much closer resemblance of humans to juvenile chimpanzees than to their elders. Our heads remain proportionally large, with brains that grow an extra third in size during our first year,

expanding beneath our flexible skulls. Our faces stay relatively flat instead of growing prominent brow ridges, and we have bodies with only the sparsest of fur. We continue to play and to learn all our lives, as though refusing, like polyps, to grow up into the more constrained lives of our ancestral ape lineage.

In the early Stone Age, we are already inveterate wanderers seeking the better life. We gather our families into protective shelters, clothe ourselves in animal skins against the cold, and use fire. We gather around fires inventing dances, songs, and finally language, becoming storytellers. Charcoal and colored muds become media for us as artists, decorating our bodies, our cave walls, our artifacts. Our powers of observation and reason develop in observing that fruit trees grow where we spit our seeds and collect kitchen wastes.

Our awe of natural powers greater than our-

180

## 150 MILLION YEARS AGO

THERMOREGULATION *Warm-blooded Dinos?*

Fossil evidence suggests that mammal-like reptiles and some dinosaurs could internally regulate their temperatures. Most contemporary reptiles, amphibians, and most fish (excepting the talented tuna) cannot do this. Birds, most mammals, skunk cabbages, and lotus flowers do regulate the temperatures of their bodies. Snakes in colder regions hibernate, lowering their metabolism to such a slow rate that even experts cannot tell if they are dead or alive. How did this regulation evolve? | Many mammals which can thermoregulate still hybernate—no sense hanging about if there's nothing to do in the heat o' the Sun nor the furious winter's rages. Is falling into deep unconsciousness, as we do at night, a pre-adaptation?

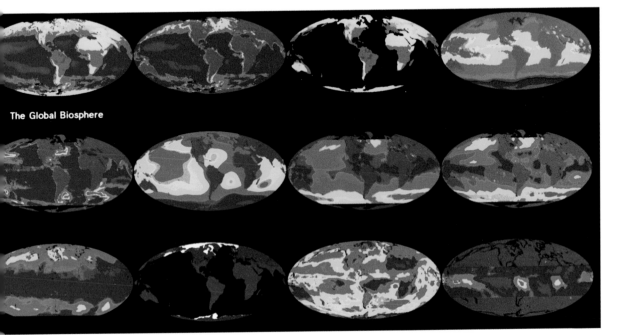

**The Global Biosphere**

The biosphere is one, biota
and environment inseparable.
Life evolves in concert with
seasonal and deep time
change. Gaian cycling is a
harmony of interchange and
modulation.

2,000 MYA                    1,000 MYA                    PRESENT

selves leads us to rituals in praise of—or to appease—powerful spirits. Plants and animals are seen as living beings like ourselves, some to be avoided and some to cherish for contributing to our well-being. We watch monkeys and other animals that learned before us which plants heal and which bring illness. Certain of us become intermediaries who communicate with other species to learn medicines and gain assistance in hunting.

During our million years of life beyond Africa, there are on the order of ten ice ages, each of them leaving wonderfully fertile watered land as the ice recedes. As our populations outgrow our lands, or when the lure of adventure seizes us, we follow the edges of the great ice sheets slowly back and forth. No doubt we encounter fellow hominids at times, but we may find each other so different or so competitive that warfare ensues. However it happened, we end up as one single species.

We capture animals for convenient food sources and make friends of some. Mostly nomads and hunters, we assemble and herd flocks, then begin to tame and ride horses. As nomads, we watch the sky for our directions, and as our imaginations expand, we project them into lightning-bolt-wielding "Father Gods" whose power demands obedience and worship. We share the fruits of our labors within our clans and

tribes; we defend them from others. Ecosystems begin to change as we move through them, hunting their animals, grazing our flocks, scattering the seeds of food we carry with us.

Some of us remain in place here and there as more settled agriculturalists, apparently tending toward the worship of a nurturing "Great Mother." We create the artifacts of settled societies through architecture and other developing arts. As nomadic tribes happen upon us, they sometimes invade and remain to rule, not infrequently "marrying" our goddess to their "Sky God," or deposing her altogether.

During and after the last ice age, we cover the whole habitable earth in our wanderings, raising towns and cities along rivers and building boats to travel and trade among them, even venturing across entire oceans. Some of us seize power over others and rule them by cunning or by force; urban class societies develop, with their armies, slaves, commerce, law, architecture, and other arts. Rulers anoint themselves as deities.

We have keen powers to observe other species and learn to daub mud into vessels and dwellings like wasps; spin thread, weave cloth, and build suspension bridges like spiders; fish like waterfowl; and burrow like moles. We even ram each others' walls like buffalo, breaking them down to plunder the riches they contain. We

182

refine our weapons and arts of war to take territory from neighbors and from cultures ever farther afield. Scratched symbols come to represent words, then alphabets are invented to write economic records, laws, military messages, and finally poetry. Now law and literature alike become visible in their own right and can be preserved unchanged across generations. Thought itself becomes an object of observation.

We begin to imagine ourselves separate from and superior to the rest of nature, seeing ourselves as its masters. Ever larger cities demand more and more surrounding "breadbasket" farming. Agriculture becomes a major undertaking as we clear rich and diverse forest ecosystems to plant single crops—monocultures of staple grains, for example—to feed burgeoning populations. Practical food economics change the human relationship with our ecosystems.

Awe and respect for provident nature gives way to our heady ability to dominate the landscape as well as each other. We have become deities who decide what shall live and what shall die. Much of the Middle East is "desertified" by huge numbers of our hoofed animals. We chop entire forests to build our cities, scrape the ground free of plants and animals we do not eat, and replace them with those we do. Without trees, mountainsides erode into river valleys,

silting into the rivers, extending their deltas. Northern Africa's coastal forests are cut by the Romans to build ships and houses, to plant wheat for bread.

Europe's forests rapidly disappear. Sometimes human populations disappear in plagues, when devastating microbes overrun crowded and unsanitary cities. We accuse each other of heresies in our attempts at religious and social conformity, torturing and burning millions of people, especially women, at the stake. At the same time, our artistic abilities soar to new heights and the wealthy live extravagantly elegant lives while the poor toil to support them.

The era of expansionist empire building begins around eight thousand years ago and becomes a dominant pattern all the way to the present. Empires come and go, tending to be warlike, though the highest philosophies of peace and harmony are created in their midst—typical of the contradictions continually expressed in our experimentally big-brained species.

Eventually, empires evolve into powerful nations with colonies spread far and wide across the seas. The printing press spreads learning; the arts flourish among the wealthy. Precious metals plundered from Europe's colonies in the Americas finance an industrial revolution that further transforms ecosystems into sprawling cities with

183

factories fueled by fossil coal.

Most humans still live simply, on the land, as traditional farmers, though indigenous populations are targeted to disappear or be enslaved under colonial rule in the name of development. White Europeans begin to dream of democracies and of the "good life" for everyone. But in practice, there is division, and wars continue to erupt until they reach global scale, turning much of our production to weaponry and the support of armies and navies.

Electricity is discovered to run machines and light cities by night. In the wake of great wars, new empires are built upon the digging out of fossil oil and its transformation into dyes, fuels, and synthetic drugs. At last having harnessed nuclear energy, we live in terror of our ability to incinerate each other with nuclear weapons that threaten friend as well as foe. Technology takes off with explosive speed, heavy industry giving way

to information-based technologies. Nevertheless, pollution mounts and ozone holes grow, our high-tech agriculture destroys soils, and rain forests and coastal waters are devastated.

We traverse the planet ever more swiftly in huge mechanical birds; we communicate instantaneously via satellites. With amazing speed, national empires morph into corporate empires that dwarf the economies of most nations. Human existence has become centered on— obsessed with—production and consumption. We are divided into a First World of affluence and a Third World where self-sustaining traditional cultures are almost gone as it feeds the First World with resources at the expense of domestic poverty.

As we have monocultured plants, we now monoculture ourselves, eliminating traditional cultures to make all our cities look alike around the world, teaching our global youth to become a uniform culture of consumers. Our medium of

---

**1 4 5  M I L L I O N   Y E A R S   A G O**

DESCENT INTO THE AIR

Flight evolves in *Archaeopteryx*; it leaps through treetops in search of insects. Birds are the only dinosaurs which will survive the coming Cretaceous-Tertiary mass extinction. Flight gives birds a leg up: they evolve long-distance seasonal migration, which permits them to responsd to Earth's hard times. | Like all migrating animals, birds have a series of preferred and backup migration systems. They read the stars for orientation; they read the landscape through sounds of water flow and low-frequency sound waves around mountain peaks and passes; they track magnetic fields by use of internal compasses.

The first known bird suggests
its dinosaur lineage.

2,000 MYA                    1,000 MYA                    PRESENT

exchange, money, takes off into speculative hyper-space, increasing the monstrous dichotomy of rich and poor. An ever smaller fraction of humanity owns most of our resources, enjoying unprecedented technological luxury, health care, and education as the bulk of our species ekes out a scant living.

The tiniest instant of time by evolutionary standards has brought us a long way from the rest of the animal kingdom in a great many ways. Yet in terms of sound ecological living, our intelligence is apparently still struggling to match that of other species. Why is it that the cleverest animal of all the large ones has caused the greatest devastation to its own natural cultures and to other species? Why do we pollute and destroy our ecosystems, even the very soils, waters, and air on which we so depend?

As we lift off from our Earth to explore our own Moon and other planets, we look back in

amazement to see our planet's fair face for the first time. And for the first time we who believed we dominate Earth seem to become truly aware of who and what we are. The Moon is as barren as it looked from here; Earth suddenly presents itself to our eyes as a lush, alive Being without political boundaries or even signs of human habitation—except for the great deserts we have made in our so very brief sojourn. We sense that it is a living home we must sustain in all its beneficent beauty.

Coming back to Earth, huge sprawling cities look strikingly like cells under a microscope. Attached to their ecosystem substrates, their structures rise against gravity, their roads and pipes and tunnels link productive facilities that generate energy and products as vehicles and people scurry about between them. They have city-center nuclei. They store and exchange information in libraries and electronic data systems, and they are

186

## 110 MILLION YEARS AGO

### DESCENT WITH CO-MODIFICATION

Earth is much more than a passive 3-D diorama or backdrop to which life-forms either adapt or go extinct. Locally, regionally, and at a global planetary level, life modifies Earth environments as much as environments shape life. Few changes have so touched vista, ecosystem, and global cycling as has the evolution of the angiosperms (plants with flowers). In the lush Cretaceous forests of fern and cycad, new colors and scents arise. Intricately woven visual and chemical communication systems set the stage for myriads of flowering plants and their animal pollinators.

188

## 65 MILLION YEARS AGO

### MEGAFAUNA MEGA-EXTINCTION

The Cretaceous period ends with a mass extinction. An asteroid six miles in diameter is believed to have hit the Yucatan Peninsula. Shock waves reverberate around Earth. Debris flying high above the atmosphere rains down with incinerating heat. Later, dust and aerosols block sunlight, and temperatures plunge. Photosynthesis stalls. All animals over 55 pounds disappear, including the dinosaurs. Many plant species disappear, and the diversity of plankton and sponges falls sharply. Approximately 85% of ocean-dwelling protoctists and marine animal species are lost. It requires 20 million years for new life-forms with high levels of diversity to reappear.

189

190

*Early primates romp on a log.*

## 55 MILLION YEARS AGO

### MAMMALS GO FORTH AND MULTIPLY

The dinosaurs were so enormous and widespread that the impact of their extinction is as great as the impact that caused it. Except for the spaces which bacteria, protoctists, and insects had inhabited, dinos had dominated Earth. | Now that dinos are extinct, the once dark and sheltered mammals stride into daylight. They move quickly to occupy available ecological niches. Among these are primates, which had evolved 30 million years earlier as forest-dwelling creatures. Primates possess several or all of the following characteristics: the ability to hold things with their hands and, sometimes, their feet; thumbs that oppose the forefinger; flattened nails in place of claws; unique teeth, skulls, and other bones; a prolonged gestation period; large brains; and acute vision with binocular

all linked across the face of Earth.

Like us, competitive ancient bacteria created cities, pollution crises, genocide, and gross inequities. But eventually they learned to share their remarkable technologies in huge cooperative multicreatured cells. Is it now up to us, among the most recent of multicelled creatures, to recognize Earth as the giant cell within which we must live together—with each other and all other species?

Already we have linked ourselves into peacefully cooperative global systems of transportation and communication, including the lightning-fast self-organizing Internet—the second great World Wide Web of information exchange. Our economics are also now globally organized, but they have created a dramatically inequitable situation for humanity and have seriously damaged ecosystems. At present we are causing the demise of other species at a rate already exceeding the onset of evolutionary extinctions.

What will it take for us to see ourselves as holons in a holarchy—as truly integral biological beings in a larger scenario than our human agenda encompasses? What will it take to recognize that we have the same opportunity to mature from competition to cooperation as any other species? Let us pause and reflect on how closely we are related not only to apelike ancestors but to the entire story of evolution, from archae to us.

## The Story of Evolution within Us

In each of us, three million potassium atoms explode every minute—far-reaching remnants of the supernova explosion that gave birth to our solar system—reminders that we are truly stardust.

In each of our trillions of cells, tens to hundreds of mitochondrial descendants of free-living

191

**capability. | Prior evolution and radiation of flowering plants—grasses, fruits and leguminous plants— provide an Eden-like world in which newly evolving mammals go forth and multiply.**

ancient bacteria labor to produce all our energy, having learned the benefits of cooperative communal life a few billion years ago.

In our developmental process as embryos, the long chain of Earth life's evolutionary stages is repeated. Biologists call it ontogeny recapitulates phylogeny—the history of the individual repeats the history of the kind. It is a living story of evolution as a prologue to our births, one that fossils cannot tell, much as they have helped us to unravel its meaning.

Consider for a moment the single fertile egg cell from which each of us develops. It is like an ancient protist formed by the sexual reproduction of its parents. It is complete in itself, yet it contains everything it needs to know to grow itself into a complex multicelled creature—in this case, a human baby. As this single cell divides again and again, it forms a hollow sphere of cells called a blastula, which resembles such protist

192

colonies as *Volvox*. This ball colony rotates as it floats in a warm salty sea of its own, created within the maternal uterine habitat and resembling the warm seas in which life began.

Then, again like its ancient forebears, it divides its labors among different groups of cells that become specialist organs. One side of the ball dents to form a groove, where the notochord develops and grows into a backbone. A lumpy head appears at one end and soon it looks like a tiny amphibian with gill slits and a twitching tail. Dark eyes bulge in its pale head.

As we keep watching, it takes on the features of other embryos in turn—those of amphibian frogs and reptilian turtles, then of avian chickens, resembling them so closely we find it hard to believe this is a human until still later. Even with arms and legs we resemble the embryos of mammals such as pigs and rabbits.

All the while we roll in our warm sea,

---

### 40 MILLION YEARS AGO
#### YOU CAN GO HOME AGAIN

Shifts in ocean currents and the development of Antarctic bottom waters drop ocean temperatures by as much as five degrees centigrade. Life thrives and expands in these nippy, nutrient-rich ocean waters. | Perhaps some combination of increased ocean productivity, shifting sea margins, the complexities of megafaunal movement on land, and a nostalgia and logic we've yet to understand—inspires the whales' land-roving mammalian ancestors, after thoughtful ambling on the edge, to return to the sea. They do go home again.

193

*Nitrogen, an element crucial for DNA, RNA, and protein syntheses, is critical for all life-forms. Grasslands and grazers spread, unmindful of their dependence upon the microcosm in Gaian cycling of nitrogen.*

## 30 MILLION YEARS AGO
### FRUITS OF EARTH

**Grasslands and trees with fruits spread. Mammals help pollinate and fertilize as they graze and swallow delicious fruits. | Earth grows cooler and more seasonal. Although extinction overtakes those animals requiring steamy tropical climates, this is largely a period of relative stability—a moment of evolutionary rest.**

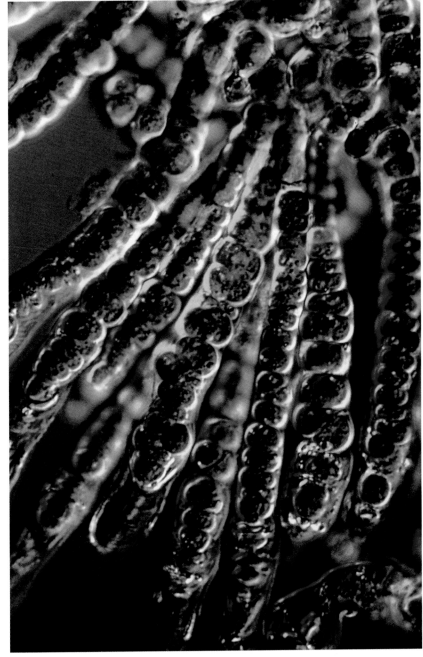

*Bacterial metabolic mastery is shown once more in this cyano's facility for fixing nitrogen. Although abundant in Earth's atmosphere, paired nitrogen atoms are loath to separate. Breaking the bond for transformation into useable form is accomplished only by lightning and bacteria.*

195

shrinking our tails as their cells voluntarily die off, as if giving in to our becoming completely human. Finally, after nine months, we are ready for birth as big-brained humans, but even then we must repeat the grand exodus from the sea amidst great maternal upheavals and gasp our first breath of air outside of our cozy marine habitat.

How amazing that this recollection of evolution is relived by each of us. How fascinating that we have developed the technology to watch our own embryonic development replay this awesome story. How marvelous to be reminded where we came from, how we are related to other species and to all our forebears back to single cells. How inspiring to know how they labored to evolve us into creatures that now have, in our own turn, such great responsibility for the survival of other species.

Seeing ourselves among them, sharing our Earth for better or worse, we wake to new

196

*Cormorants are among the most ancient of living diving birds. With special retractable lenses, their underwater vision is unsurpassed.*

## 20 MILLION YEARS AGO

### PRESSURES MOUNT

Tectonic pressures mount and mountain ranges form—the Cordilleras, the Andes, the great Himalayan range. As inland seas shrink, the climate wavers through extremes of hot and cold.

Ocean currents change and nutrients well up from the deep, supporting enormous growth of photoplankton. This foundation of the ocean food web prospers, allowing other species to prosper as well.

Seals and sea lions flourish. Species of diving birds diversify. With falling sea levels, land bridges connect Siberia to North America and England to the continent. Grand parades of intercontinen-

tal migrations take place. Most great groups of mammals (mammalian orders) appear essentially modern in form. Fossils from the state of Nebraska include camels, deerlike animals, bear,

197

*The ancestry of modern great apes is not well known. Proconsul, an early member of the hominid lineage, may have been too primitive in a number of respects to represent a link in the evolutionary chain to modern forms.*

dogs, foxes, peccaries, small beavers, ground squirrels, and horses. Horse evolution occurs primarily in North America. Early elephantlike creatures spread from Africa to the Eurasian continent.

possibilities. Just as they created crises, we create them in our turn. Just as they evolved through cooperative solutions, we can be inspired by them to do the same.

We have called ourselves Homo Sapiens— Self-aware Man, Thinking Man, Intelligent Man, or Wise Man, depending on interpretation. Whichever meaning we prefer, we stand today at a crossroads where our self-assumed name is challenged. Will we continue to cause a major extinction? Will we commit species suicide through nuclear holocaust, germ warfare, global warming, or poisoned ecosystems, as leading scientists warn? Or will we mature in the face of crisis as our forebears did? Will we live up to our name as wise beings, ending our destruction of ecosystems, waters, and atmosphere; our gross economic

inequities; our devotion to weapons as ways to solve conflict? Will we become as wise as the ancient bacteria in cooperating to move evolution to a higher stage where all benefit?

The Etruscan root of sapiens is closer in literal meaning to "being enlightened." What a wonderful meaning to realize at this most challenging and fascinating of evolutionary moments.

## 10 MILLION YEARS AGO
### GETTING FAMILIAR

Breakthroughs in molecular biology are revolutionizing our understanding of the history and relationships of hominid primates. Orangutans depart from the combined African great ape and human line about 13 MYA. The great ape (or gorilla) lineage splits from the combined chimpanzee and human lineage some 8 MYA. We share 99% of our DNA with chimps. This does not mean that we are "descended" from chimps but rather that we last shared a common ancestor 8 MYA.

*The Indian species* Ramapithecus punjabicus, *an ape speculatively reconstructed here from fossil evidence, belongs to our family* Hominidae.

199

**5 - 0** **MILLION YEARS AGO**

UP TO US

Chimpanzee and human
lines split 5 MYA.
Humans and chimps
share close to 99%
of genes.

Humans build shelters
and use increasingly
refined tools, weapons,
fireplaces, grease
lamps, and sewn pelts.

200

*Homo erectus* emerges
from Africa.

*Neanderthal* man coexists
with *Cro-Magnon* man in
Western Europe

*Australopithecus*
demonstrates hominid
bipedalism.

*Archaic sapiens* is one of the
oldest fossil humans found in
Europe.

Stone tools appear, possibly
associated with *Homo habilis.*

4 MYA

3 MYA

2 MYA

MYA = million

***Cro-Magnon*** man returns from a hunt in the Dordogne region of France.

A Late Stone Age European artist crafts a figurine. *Homo sapiens sapiens* spread over the world, entering North America over the Bering land bridge.

Many large North American animals, such as this woolly rhino, become extinct, perhaps due to overhunting by humans.

201

Early European *Homo sapiens sapiens* kill a mammoth. Three different species of humans coexist: Neanderthal man (Homo sapiens neanderthalensis) in Asia and eastern Europe, *Homo erectus* in Asia, and the relentlessly spreading *Homo sapiens sapiens*.

And, today, many scientists believe we are in the midst of another mass species extinction, caused by humans, which threatens Earth's glorious biodiversity.

Human Population (billions)

6

5

4

3

2

1

0

1 MYA

Present

# LOOKING TO THE FUTURE

**The Sun can continue to support life on Earth for 2,000 to 3,000 million years (2,000 to 3,000 feet) into the future.**

**What will you do, what will we do, to help preserve, to help realize the possibilities?**

202

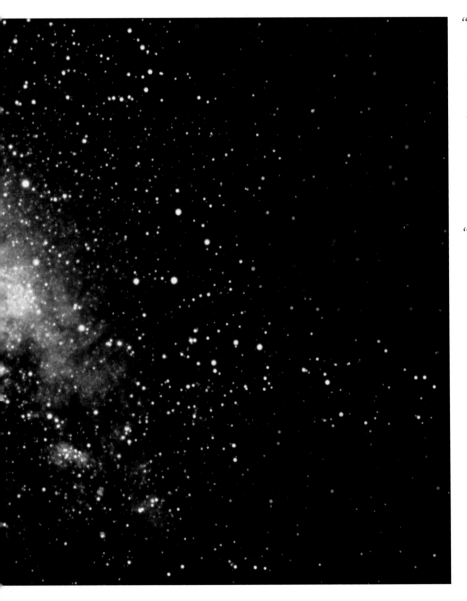

*"No one can predict the future. All we can do is choose our contribution to the circumstances which, perhaps, will help shape the future."*

Don Michael

203

*"In wildness is the preservation of the world."*

Henry David Thoreau

# EPILOGUE:
## CRISIS AND OPPORTUNITY

*The real voyage of discovery does not consist of seeking new landscapes, but in having new eyes.*

Marcel Proust

The story of our origins revealed by modern science is overwhelmingly beautiful, awe-inspiring, and humbling. Photons of radiant energy issue forth in the Big Bang. They, in turn, transmute into fundamental particles, which coalesce into hydrogen nuclei and first generation stars. These stars cook up the heavier elements, then explode as supernova, spewing stardust into space to recondense into second generation stars and planetary systems. A succession of increasingly complex molecular organizations evolve into life as we know it on this precious planet Earth, issuing forth each of us to reflect on the wonder of it all.

We have seen that we, along with all other living beings, are the product of an astounding 15-billion-year evolutionary process, that every living being is related to every other living being, that our bacterial ancestors created the oxygen of our atmosphere, invented every metabolic process known to modern science, contributed to the iron ore deposits that sustain our technological society, and evolved into the very cells of our bodies. While competition and predation play significant roles in life, cooperation and mutual support among species is a dominant and recurring theme.

How is it, then, given our genesis, that we so commonly treat our planet and one another so poorly and fail adequately to address the problems of species extinction, wilderness habitat destruction, rain forest burning, ozone hole expansion, carbon dioxide green house build up, and human overpopulation? Is it that we are ignorant of the facts? Do we take it all for granted? Are we genetically incompetent to behave otherwise? Or is there some more profound explanation? Is there hope for something better?

We opened our preface with a quote from author Francis Moore Lappe: "Each of us carries within us a worldview, a set of assumptions about how the world works—what some call a paradigm—that forms the very questions we allow ourselves to ask and determines our view of future possibilities." The nineteenth century philosopher Arthur Schopenhauer put it succinctly. "Every man takes the limits of his own field of vision for the limits of the world."

Is it possible that the prevailing current worldview lacks long-term survival value and is in urgent need of renewal? This book is motivated by the belief that the answer is yes, and that knowledge of the evolution of the universe and life as we know it constitute a fundamental

204

context for identifying and addressing the most critical issues of the present and the future, and for inspiring a grander vision.

There is growing worldwide realization of, and concern for, the magnitude of human impact on the planet. Modern societies tend to view the world as an expendable resource, available for human consumption to serve short-term purposes, with minimal consideration of the impact on the near and long-term future of life. Our pattern of exhausting non-renewable resources and destroying ecological capital is sustainable for but a moment in time. We fail to consider that there will be times of crisis in the future when readily accessible fossil, mineral, and biological resources will be of immensely greater value to society. Concern for the future would dictate preservation of non-renewable resources rather than their profligate consumption.

It is a further tragedy that we continue to wipe out the diversity and wisdom of indigenous cultures. These peoples, while not always treating kindly tribes with whom they competed for resources, appear generally to have manifested a deep knowledge and appreciation of, and a reverence for, nature. Their knowledge and wisdom could have served humanity well in the future.

Given that the Sun can support life on Earth for at least another couple of billion years,

prudence would suggest that we strive to identify, and then to avoid taking, the most significant irreversible actions. Some human-precipitated environmental insults might be ameliorated in a matter of years, decades, or centuries. Examples could be healing the hole in the ozone layer, reducing the planet's burden of heat-trapping carbon dioxide, or reducing and then ceasing the consumption of non-renewable resources. Other critical injuries to the very fabric of life require over a thousand times as long to heal. The fossil record indicates that Earth has experienced five mass extinctions of species in the past 500 million years. The record, further, reveals that the time required for biodiversity levels to recover is measured in *tens of millions of years*.

Only a small fraction of the public is aware that humans are precipitating what could rapidly become the greatest mass extinction of species in the past 65 million years, having irreversible impact on the viability of the ecosystem and quality of life on Earth for tens of millions of years into the future. A recent survey commissioned by The American Museum of Natural History in New York City established that biologists view the gravity of the extinction crisis to surpass that of pollution, global warming and thinning of the ozone layer.

A principal precipitating factor in the

205

pending mass extinction is human degradation and destruction of habitat. Few conservation biologists would argue that it is possible for larger wild animals to evolve for millennia into the future confined to isolated few-thousand square mile wildlife preserves surrounded by dense populations of humans. According to population geneticists, the population sizes of large animals in these preserves is too small for natural selection to operate effectively. A one-thousand-square-mile park averages only some thirty miles across. In even the largest of parks, creatures cannot maintain requisite genetic diversity, or escape spread of disease. The well-publicized destruction of rain forests, home to most of the land species of animals and plants, continues unabated.

Some people may feel that the extinction of a large fraction of species of life on Earth is not a significant issue, provided that these extinctions do not precipitate, as they may well do, the demise of domestic plants and animals, and ourselves. But can it really be that the majority of us, when informed, will argue with comfort and confidence that we should be party to the process of causing a mass extinction, from which it would take tens of millions of years to recover, without first exhaustively exploring the possible ramifications and issues in informed comprehensive global debate?

For some of us the most compelling arguments are premised on spiritual and moral grounds. We cannot accept being party to a process of extinguishing the wonder and beauty of life that has been 15 billion years in the making and could continue evolving far into the future, nor standing idly by while others do so. We do not feel qualified to assess the relative merit of existence of different species.

The actions required to slow and stop the current extinction will require a dramatic shift in attitude and behavior. Such a shift will favor an enhanced quality of life for humans as well as for other life, and favor the long-term survival of humanity. Changing our attitudes will be hard work, requiring an enhanced appreciation and respect for the natural processes that sustain life.

Critical among these actions would be a gradual substantial reduction in the size of the world's human population. The opportunity to live in the most naturally habitable places in the world, in tranquility, near nature, free from crowding, din, and congestion, is for many of us, a fundamental requirement for quality existence. These qualities are eroding daily as human populations swell and inundate the formerly most liveable regions on Earth. In addition, human overpopulation of the planet is the principal precipitating factor in the destruction of habitat.

An average of two or slightly fewer children per child-bearing couple would enable substantial reduction of the world's population over the next few hundred years, allowing the gradual but significant enlargement and reconnection of wilderness areas.

Though this perspective may seem severe, we must realize that averting the pending extinction crisis requires urgent, dramatic, and massive response. *The Earth is losing an estimated three or more species per hour, a rate one hundred to one thousand times greater than the average over the preceding hundreds of millennia. And the trend accelerates. We have lost over ten percent of the species that were living a few hundred years ago. Conservation biologists are predicting that half of the diversity of life will be lost in the next century if the present rates of habitat destruction and disturbance continue.*

Conservation biologists argue that wild areas need to be expanded and connected in order to preserve wild species. Wild populations of plants and animals have tremendous capacity to recover and expand, given viable populations, and the space to do so. Until proven otherwise prudence would suggest heeding Henry David Thoreau's dictum "In wildness is the preservation of the world." For it is in 4,000 million years of wildness that all living species evolved. It is wildness that created the spectacular life system flourishing on

this planet, creatures that replicate, self-repair, support one another, wonder about the universe, and evolve into the future. It has not been established that human-guided evolution of life on Earth can and should substitute for the robust and exquisite process that brought us this far.

The Golden Rule, in one form or another, is almost universal among the religions of the world, be it expressed either in the affirmative, as "Do unto others as you would have them do unto you," or in the negative, "Do not do unto others what you would not have them do unto you." To the extent that the rule has been practiced as well as preached, its application has been restricted to humans, and in particular those with whom one comes directly into contact. This was reasonable in the past, at a time when our impacts tended to be local in space and in time. However, human conduct today affects the entire planet, for millions of years into the future. Consequently, it is timely to extend the Golden Rule—to all life, over the entire planet, and as far into the future as our behavior will have significant impact.

An implicit thesis of this book is that science and technology have much to contribute toward achievement of that wisdom. Science and derivative technology forge a double-edged sword. On the one hand, scientific knowledge enables

208

technologies for transforming raw material into tools and devices, developing processes for their application, fulfilling our material desires, extending our creative abilities, easing the physical tasks of survival, enhancing the likelihood of living to old age, in short granting us immense power. When used unwisely that power jeopardizes the future of all life.

On the other hand, science and technology expand our horizons, extending our sensory capacities to observe the large and the small in the universe, yielding theories to organize our perceptions and to predict previously unknown phenomena, and to reveal to us spectacular details of new domains of existence. Science and technology contribute fundamentally to our understanding of our place in space and time, and our origins. The knowledge, context and perspective offered to us by modern science and technology have an unrealized potential to make a major contribution to a guiding wisdom commensurate with our power to affect the present and the remote future.

A long-term future-oriented worldview would keep us well back from precipices, and obviate last-minute debates on the likelihood, or timing, of our falling off. The wonder of the universe, life, and conscious self-reflection, of love, caring and spirituality enfold us so intimately that we largely fail to see it.

Is it possible that a sense of awe, wonder and humility, of origins, place, possibilities, and recovery of a belief in the sacredness of nature, can, and perhaps must, become operational imperatives in guiding humanity into the future? Rather than pondering the illusive purpose of life, can we not accept and appreciate the gift, live the life we are given, respect all life, and preserve options for the future. Though none of us has the power to control the future, each of us is free to determine how we will contribute to the circumstances out of which the future will evolve.

Sidney Liebes

# REFERENCES

*References to Text by Elisabet Sahtouris:*

1   Monastersky, Richard. "The Rise of Life on Earth." *National Geographic* vol. 193, no. 3, March 1998.

2   Thomas, Lewis. *The Lives of a Cell: Notes of a Biology Watcher.* Bantam: New York, 1975.

3   Lovelock, James. *Gaia: A New Look at Life on Earth.* Oxford: Oxford University Press, 1995.

4   Lovelock, James. *The Ages of Gaia: A Biography of Our Living Earth.* New York: W.W. Norton, 1988.

5   Schneider, S. H. & Boston, P. *Scientists on Gaia.* Cambridge, MA: The MIT Press, 1993.

6   Fyfe, W. S. "The Biosphere Is Going Deep." *Science,* Vol. 273, 2226, July 1996.

7   Fyfe, W. S. *Handbook of Environmental Chemistry.* New York: Springer-Verlag, 1994.

8   Lapo, A. V. *Traces of Bygone Biospheres.* Mir Publishers, 1982.

9   Vernadsky, Vladimir I. *The Biosphere.* New York: Copernicus, 1997 (originally published in the Soviet Union in 1926).

10   Maturana, Humberto R. and Francisco Varela. *The Tree of Knowledge: The Biological Roots of Human Understanding.* Boston: Shambala, 1987.

11   Capra, Fritjof. *The Web of Life.* New York: Doubleday Anchor, 1996.

12   Sahtouris, Elisabet. *Earthdance: Living Systems in Evolution.* Santa Barbara, CA: Metalog, 1995 (available at HYPERLINK http://www.ratical.com/lifeweb www.ratical.com/lifeweb).

13   Ho, Mae-Wan and Fox, S.W., eds. *Evolutionary Processes and Metaphors.* London: Wiley, 1988.

14   Lipton, Bruce. "The Biology of Consciousness." Ft. Collins, CO: Proc. of the Int'l Assoc. of New Sciences, 1993.

15   Sonea, S. and M. Panisset. *A New Bacteriology.* Boston: Jones & Bartlett, 1983.

16   Margulis, Lynn. *Symbiosis in Cell Evolution: Microbial Communities in the Archean and Proterozoic Eons.* New York: W. H. Freeman, (2nd edition), 1993.

17   Margulis, Lynn. *Early Life.* Boston: Jones & Bartlett, 1984 (earlier edition Boston: Science Books International, 1982).

18   Margulis, L. and Sagan, D. *Microcosmos: Four Billion Years of Evolution from our Microbial Ancestors.* London: Allen & Unwin, 1987.

19   Bloom, Howard. *Global Brain: The Evolution of Mass Mind from the Big Bang to the 21st Century.* New York: John Wiley & Sons, (in preparation).

20   Coghlan, Andy. *New Scientist.* London: Aug. 31, 1996.

21   Coghlan, Andy. *World Press Review.* December 1996, pp. 32-33.

22   Shapiro, James A. and Dworkin, Martin, eds. *Bacteria as Multicellular Organisms.* Oxford: Oxford University Press, 1998.

23   Ben-Jacob, Eshel. "From snowflake formation to growth of bacterial colonies II: Cooperative formation of complex colonial patterns." *Contemporary Physics,* 1997 vol. 38, no. 3, pp. 205-241.

24   Ben-Jacob, Eshel. "Bacterial wisdom, Goedel's theorem and creative genomic webs." *Physica* A 24, 199, pp. 57-76.

209

25 Friedman, Norman *The Hidden Domain: Home of the Quantum Wave Function. Nature's Creative Force.* Eugene, OR: The Woodbridge Group, 1997.

26 Keller, E. F. *A Feeling for the Organism: The Life and Work of Barbara McClintock.* San Francisco: Freeman, 1983.

27 McClintock, Barbara. "The significance of responses of the genome to challenge." *Science* 226, pp. 792-801, 1984.

28 Sharp, P. A. "Split genes and RNA splicing." *Cell* 77, pp. 805-815, 1994.

29 Shapiro, J. A. "Natural genetic engineering in evolution." *Genetica* 86, pp. 99-111, 1992.

30 Cairns, J. J. Overbaugh and S. Miller. "The Origin of Mutants." *Nature* 335, pp. 142-145, 1988.

31 Rasicella, J. P., P. U. Park and M. S. Fox. "Adaptive Mutation in Escherichia coli: A Role for Conjugation." *Science* 268, pp. 418-420, 199.

32 Temin, H. M., and W. Engels. "Movable Genetic Elements and Evolution." in J. W. Pollard, ed. *Evolutionary Theory: Paths into the Future.* Chichester: John Wiley & Sons, 1984.

33 Pollard, Jeffrey. "New Genetic Mechanisms and Their Implications for the Formation of New Species." in Ho, Mae-Wan and Sidney Fox, eds. *Evolutionary Processes and Metaphors.* Chichester: John Wiley & Sons.

34 Sagan, Dorion and Lynn Margulis. *Garden of Microbial Delights: A Practical Guide to the Subvisible World.* New York: Harcourt Brace Jovanovich, 1988.

35 Ingber, Donald E. "The Architecture of Life." *Scientific American.* January 1998.

36 Glantz, James. "Force-carrying Web Pervades Living Cell." *Science,* vol. 276, 2 May 1997.

37 Maniotis, A. J., K. Bojanowski and D. Ingber. "Mechanical continuity and reversible chromosome disassembly within intact genomes removed from living cells." *Journal of Cellular Biochemistry,* 65, pp. 114-130.

38 Margulis, Lynn and Dorion Sagan. *What Is Life?* New York: Simon & Schuster, 1995.

39 Monastersky, Richard. "The Rise of Life on Earth: Life Grows Up." *National Geographic,* vol. 193, no. 4, April 1998.

40 Article on pre-Cambrian Chinese phosphate mine fossils in *Nature,* Feb. 5th 1998.

41 Gould, Stephen J. *Ever Since Darwin: Reflections in Natural History.* New York: Penguin, 1977.

42 Gould, Stephen J. *The Mismeasure of Man.* New York: Penguin, 1984.

43 Gould, Stephen Jay. *Wonderful Life: The Burgess Shale and the Nature of History.* New York: W. W. Norton, 1989.

44 Vogel, Gretchen. "Early Start for Plant-Insect Dance." *Science,* Vol.273, August 16 1996, p. 872.

45 Fastovsky, David E. and David B. Weishampel. *The Evolution and Extinction of the Dinosaurs.* Cambridge: Cambridge University Press, 1990.

46 Chadwick, Douglas H. "Planet of the Beetles." *National Geographic,* vol. 193, No. 3, March 1998.

47 Popescu, *Petru Amazon Beaming.*

48 Watson, Lyall. *Beyond Supernature.* Bantam: New York, 1988.

49 Colbert, Edwin H. *Evolution of Vertebrates: A History of the Backboned Animals through Time.* New York: John Wiley & Sons, 1988.

210

50   *Paleoworld: Evolution of Whales.* Documentary by New Dominion Pictures for The Learning Channel, 1994.

51   Cousteau, Jacques, and Yves Paccalet. *Whales.* New York: Harry N. Abrams, 1988.

52   Harrison, Richard, and M. M. Bryden, eds. *Whales, Dolphins, and Porpoises.* New York, Oxford, and Sydney: Golden Press Pty Ltd., 1988.

53   May, John, ed. *The Greenpeace Book of Dolphins.* New York: Sterling Publishing Company, and London: Century Editions, 1990.

54   Koestler, Arthur 1978. *Janus: A Summing Up.* London: Pan Books.

55   Wilber, Ken. *A Brief History of Everything.* Boston : Shambhala, 1995.

56   Harman, Willis and Elisabet Sahtouris. *Biology Revisioned.* Berkeley: North Atlantic Publishers, 1998.

57   Sahtouris, Elisabet. "The Biology of Globalization." in *Perspectives in Business and Global Change,* Vol. 11, No. 3, 1997, World Business Academy.

### Additional Readings:

Abram, David. *Spell of the Sensuous.* New York: Pantheon Books, 1996.

Benyus, Janine. *Biomimicry: Innovation Inspired by Nature.* New York: William Morrow & Co., 1997.

Berry, Thomas. *The Dream of the Earth.* San Francisco: Sierra Club Books, 1988.

Darnell, J. et al. *Molecular Cell Biology.* New York: Scientific American Books, Inc., third edition, distributed by W. H. Freeman and Co., 1995.

Hutchinson, G. E. *The Ecological Theater and the Evolutionary Play.* New Haven, CT: Yale University Press, 1965.

Jantsch, Erich. *The Self-Organizing Universe.* Oxford: Pergamon Press, 1980.

Keller, E. F. *A Feeling for the Organism: The Life and Work of Barbara McClintock.* San Francisco: Freeman, 1983

Odum, Eugene P. *Basic Ecology.* Philadelphia: Saunders College Publishing, 1983.

Swimme, Brian and Thomas Berry. *The Universe Story.* San Francisco: Harper, San Francisco, 1994.

### Bibliography for Exhibition Text:

Bengtson, Stefan, ed. *Early Life on Earth,* Nobel Symposium No. 84. New York: Columbia University Press, 1994.

Cowen, Richard. *History of Life.* Cambridge, MA: University of California, Davis, Blackwell Scientific Publications (second edition), 1995.

Gould, Stephen Jay. *The Book of Life.* New York: W. W. Norton Company, 1993.

———. *Wonderful Life: The Burgess Shale and the Nature of History.* New York: W.W. Norton, 1989.

Eldredge, Niles. *Fossils: The Evolution and Extinction of Species.* New York: Harry N. Abrams, Inc., 1991.

Lovelock, James. *Ages of Gain: A Biography of Our Living Earth.* New York: W.W. Norton, 1994.

Lowenstam, Heinz and Stephen Weiner. *On Biomineralization.* Oxford: Oxford University Press, 1989.

Margulis, Lynn. *Early Life.* Boston: Jones & Bartlett, 1984.

211

——. *Symbiosis in Cell Evolution: Microbial Communities in the Archean and Proterozoic Eons.* New York: W. H. Freeman, (2nd edition), 1993.

Margulis, Lynn and Lorraine Olendzenski, eds. *Environmental Evolution: Effects of the Origin and Evolution of Life on Planet Earth.* Cambridge and London: MIT Press, 1994.

Margulis, Lynn and Dorion Sagan. *Microcosmos: Four Billion Years of Evolution from our Microbial Ancestors.* Berkeley: University of California Press, 1997.

——. *Garden of Microbial Delights: A Practical Guide to the Subvisible World.* Dubuque, IA: Kendall/Hunt, 1993.

——. *Origins of Sex: Three Billion Years of Genetic Recombination.* New Haven, CT: Yale University Press, 1991.

——. *What Is Life?* New York: Simon & Schuster, 1995.

McMenamin, Mark and Dianna McMenamin. *Hypersea: Life on Land.* New York: Columbia University Press, 1989.

Schneider, Stephen. N. and Penelope Boston, eds. *Scientists on Gaia.* Cambridge, MA: MIT Press, 1993.

Vernadsky, Vladimir I. *The Biosphere.* New York: Copernicus, 1997.

Westbroek, Peter. *Life as a Geological Force.* New York and London: W. W. Norton, 1992.

## Complementary References and Recommended Reading:

*Britannica OnLine.* Encyclopaedia Britannica, Inc., 1994–1997.

Mazak, Vratislav. *Prehistoric Man: The Dawn of Our Species.* London: The Hamlyn Publishing Group

Limited, 1980.

Schopf, William J. *Major Events in the History of Life.* Boston, MA: Jones and Bartlett Publishers, 1992.

Spinar, Zendak V. *Life Before Man.* London: Thames and Hudson (second edition), 1995.

Spinar, Zdenek V. and Dr. Philip J. Currie. *The Great Dinosaurs, A Story of the Giants' Evolution.* Stamford, CT: Longmeadow Press, 1994

Wilson, Edward O. *The Diversity of Life.* Cambridge, MA: The Belknap Press of Harvard University Press, 1992.

## Further Suggested Reading:

Barrow, John D. *The Origin of the Universe.* New York: Basic Books, 1994.

Couper, Heather and Nigel Henbest, and illustrated by Luciano Corbella. *Big Bang: The Story of the Universe.* New York: Dorling Kindersley Publishing, 1997.

Couper, Heather and Nigel Henbest. *The Guide to the Galaxy.* New York: Cambridge University Press, 1994.

Smolin, Lee. *The Life of the Cosmos.* New York: Oxford University Press, 1997.

Volk, Tyler. *Gaia's Body: Towards a Physiology of Earth.* New York: Springer Verlag, 1998.

# Glossary

**accrete** *to grow or enlarge by gradual buildup*

**acid** *substance that in water solution tastes sour; reacts with base to form a salt*

**aerobic** *living in the presence of oxygen*

**algae** *nonvascular aquatic ancestors of plants, often resembling plant forms*

**amino acids** *organic compound constituents of proteins, composed of carbon, hydrogen, oxygen, and nitrogen and sometimes sulfur*

**amphibian** *vertebrates able to exploit both aquatic and terrestrial environments (from the Greek* amphibios *"living a double life")*

**anaerobic** *living in the absence of oxygen*

**anastomose** *to connect or join together (as do streams, or leaf veins, etc.)*

**animal** *multicellular eukaryote organism that develops from an embryonic blastula*

**anti-particle** *a fundamental particle having negative mass and opposite charge to the particle*

**apoptosis** *a mechanism that allows cells to self-destruct when stimulated by the appropriate trigger; "programmed cell death"*

**archaea** *an early prokaryotic microbe, distinct from bacteria, that first appeared nearly 4 billion years ago*

**Archean** *the first three billion years plus of life's history, up to the Paleozoic*

**arthropod** *animals, including lobsters, crabs, spiders, and insects, with exoskeleton and segmented body to which internal muscles are attached*

**ATP** *adenosine triphosphate, the metabolic "energy currency" of all cells*

**asteroid** *any of the small celestial bodies found generally between the orbits of Mars and Jupiter*

**autopoiesis** *self-perpetuation through metabolization, involving consumption of energy and discard of waste (from Greek self [auto] and making [poiein, as in "poetry"]); active maintenance against natural degradation*

**autotroph** *any organism that obtains carbon directly from carbon dioxide ($CO_2$)*

**bacterium (plural: bacteria)** *unicellular microscopic prokaryote, also called moneran*

**banded iron formation (BIF)** *old sedimentary rock comprising alternating layers of more or less oxidized iron oxides*

**base** *substance that in water solution tastes bitter, and is slippery to the touch; reacts with acid to form a salt*

**benthic** *occurring in the depths of the ocean*

**BIF** *see* banded iron formation

**Big Bang** *inferred event of explosive emergence of the known universe from a state of extremely high energy and temperature*

**billion** *one thousand million*

**biology** *science of living organisms*

**biomineralization** *the concentration by an organism of a mineral compound*

**biosphere** *living beings and their environment*

**biota** *the flora and fauna of a region*

**biped** *two-footed animal*

**blastula** *hollow sphere of cells produced during the development of an embryo by repeated cleavage of a fertilized egg*

**bluegreen** *popular name given, due to their color, to photosynthesizing cyanobacteria*

**boson** *a particle (as a photon or meson) whose spin is either zero or an integral number*

**breathers** *fermentors that live by partially breaking down ready-made food molecules and emitting product gases*

**bubbler** *popular name given to fermenting bacteria because of their gas emission*

213

**Burgess Shale**  *Western Canadian site of Cambrian fossils*

**cartilage**  *translucent, plasticlike component of the skeletons of certain primitive vertebrates and mammalian embryos*

**centromere**  *cell organelle that holds pairs of chromosomes together during reproductive phase*

**cetacean**  *collective name for several species of aquatic mammals including whales, porpoises, and dolphins*

**chemistry**  *a science that deals with the composition, structure, and properties of substances and with their transformation*

**chemoautotroph**  *life-form living off chemical oxidation reactions and carbon dioxide*

**chemoheterotroph**  *life-form living off chemical oxidation reactions and complex compounds*

**chert**  *smooth form of black quartz*

**chimera**  *an illusion or fabrication of the mind*

**chlorophyll**  *green chemical compound participating in photosynthesis*

**chloroplast**  *cell organelles descended from cyanobacteria, active in photosynthesis*

**chordate**  *a member of the phylum Chordata*

**Chordata**  *phylum (of which vertebrates are a subphylum) of animals containing, at some time in their life cycles, a notochord stiffening rod*

**chromosome**  *eukroytic chromosomes are "packaged" versions of gene-carrying DNA intricately folded around proteins, along with some RNA; prokaroyotic chromosomes consist entirely of DNA*

**cilia**  *eyelash-like; waving or rotating "hairs" that move protists and plant sperm and perform diverse other functions*

**clone**  *genetically identical replica*

**codon**  *a unique sequence of three of the four bases adenine, thymine, cytosine, and guanine in a nucleic acid chain that encodes for specific amino acid*

**combustion**  *burning*

**comet**  *small celestial object orbiting the Sun that develops diffuse gaseous envelopes and often long luminous tales when near the Sun*

**coprolite**  *fossilized dung*

**cosmos**  *in astronomy, the entire physical universe, consisting of all objects and phenomena observed or postulated*

**Cro-Magnon**  *anatomically modern Homo Sapiens living during the period 35,000 to 10,000 years ago*

**crust**  *the outermost solid part of Earth*

**cryptocrust**  *the invisible crust of bacteria blanketing Earth*

**cyanobacterium**  *a prokaryotic photosynthesizing life-form*

**cyst**  *a resting stage formed by some bacteria and protozoa wherein entire cell is surrounded by a protective layer*

**deep sea vent**  *ocean floor fissure through which magma and/or chemically rich hot water pours*

**deep-time**  *time in the biological and geological distant past (analogous to astronomical "deep-space")*

**diatom**  *any of about 16,000 species of eukaryotic protist forming elaborate shells*

**diploid**  *in eukaryotes, an organism with two chromosome complements, one derived from each haploid gamete*

**DNA**  *deoxyribonucleic acid; double helix molecule arranged in genetic code*

**Ediacaran biota**  *Precambrian fossils found in the Ediacara Hills, north of Adelaide, South Australia*

**electromagnetic radiation**  *intimately coupled waves of electric and magnetic fields that propagate with the speed of light; examples are visible light and radio waves*

**electron**  *lightest known stable subatomic particle, "clouds" of which surround the nuclei of an atom*

**elementary particle**  *subatomic particle*

**endosymbiosis**  *symbiosis with symbiont dwelling within the body of its symbiotic partner*

**enzyme**  *highly active protein that catalyzes biochemical reactions within cells*

214

**eon (geological)**  *the largest unit of geological time*

**epoch (geological)**  *subdivision of a geological period of time*

**era (geological)**  *subdivision of a geological eon of time*

**eubacteria**  *one of two major groups of prokaryotes; the other: archaebacteria*

**eukaryote**  *(pronounced you-CARRY-ote) cell containing a clearly defined nucleus*

**evolution (biology)**  *modification in successive generations of organisms*

**exoskeleton**  *rigid or articulated envelope that supports and protects the soft tissues of certain animals, such as lobsters, crabs, spiders, and insects*

**expressed genes**  *the relatively small fraction of genes that significantly influence the gene product*

**family (taxanomic)**  *taxonomic group in the hierarchy: kingdom, phylum, class, order, family, genus, species*

**fauna**  *animal life*

**feedback**  *control of a biological reaction by the end products of that reaction*

**fermentation**  *metabolic process involving anaerobic breakdown of sugars into lactic acid or into carbon dioxide and alcohol*

**flagellum (plural: flagella or flagellums)**  *spinning corkscrew-shaped locomotive appendage of some cells (see proton motor)*

**flora**  *plant life*

**foram**  *(abbreviation; see foraminifer)*

**foraminifer (plural: foraminifera)**  *type of protist having calcareous shell*

**fossil**  *remnant, impression, or trace of an organism of past geologic age preserved in Earth's crust*

**fungi (singular: fungus)**  *eukaryotes such as yeasts, rusts, smuts, molds, mushrooms, and mildews*

**future**  *in the context of this book, the potential 2,000–3,000 million years that the Sun can continue to support life on Earth*

**Gaia**  *James Lovelock's designation for active biological control of planetary environments by life, or for the coupling of biotic (life) and abiotic (nonife) activity*

**galaxy**  *one of nearly a million million large collections of stars each typically containing nearly a million million stars*

**gamete**  *reproductive cell, such as egg, sperm, pollen*

**gene**  *unit of hereditary information on a chromosome, usually defined as the code for a single protein*

**genome**  *the complete genetic material of an organism*

**geodesics**  *Buckminster Fuller's term for his architecturally geometrical tensegrity structures, mainly domes*

**geology**  *a science that deals with the history of Earth and its life especially as recorded in rocks*

**greenhouse effect**  *warming of Earth caused by increased concentrations of radiant-heat-trapping atmospheric gases such as carbon dioxide*

**Hadean eon**  *period of Earth history from about 4,600 to 3,900 million years ago*

**haploid**  *half complement of chromosomes in a sex cell before union with another*

**heterotroph**  *life-form living off complex chemical compounds*

**homeostasis**  *balanced fluctuations around a single reference point*

**hominid**  *one of a family (Hominidae) of erect bipedal primate mammals including recent humans*

***Homo erectus***  *speculated to be direct ancestor to human, living from 1,600,000 to 250,000 years ago*

***Homo sapiens***  *human ancestors appearing some 400,000 years ago*

***Homo sapiens sapiens***  *modern humans*

**hydrosphere**  *the aggregate of planetary waters, both sweet and saline*

215

**hypersea** *the movement of seawater onto land inside organisms*

**inorganic** *any compound composed of two or more chemical elements other than carbon*

**intranet** *internal network*

**kingdom** *one of five life groups: Protictista, Plantae, Animalia, Fungi, Monera*

**legumes** *dry fruits such as peas, beans, and vetch*

**life** *there is no generally accepted definition of life; conventionally viewed as the quality that distinguishes a vital functional being from a dead body (see also* autopoiesis)

**liposome** *tiny, fatty, bubblelike capsule*

**lithosphere** *the outer rocky crust of Earth*

**macrocosmos** *realm of the very large*

**macroscopic** *relatively large (not requiring microscope to see)*

**mammal** *animal giving birth to live young and nurturing them on milk*

**medusae** *offspring of polyps, budding from them and swimming as fringed jellyfish*

**meiosis** *cell division without chromosome duplication, resulting in offspring with haploid chromosomes*

**metabolism** *the biochemical processes of living entities; sum of anabolic buildup and catabolic breakdown of compounds and cycles*

**metamorphose** *to change into a different physical form*

**meteor** *a streak of light in the sky that forms when a small chunk of rock or metallic matter enters Earth's atmosphere and vaporizes; also, the particle itself*

**meteorite** *a chunk of stony or metallic matter that survives flight from outer space through Earth's atmosphere and lands on the ground*

**microbe** *microscopic life-form*

**microbiology** *branch of biology dealing with microscopic life-forms*

**microcosmos** *microscopic or submicroscopic world*

**micrometer** *one millionth of a meter; sometimes abbreviated "micron"*

**microscopic** *tiny (generally requiring microscope to see)*

**million** *one thousand thousand*

**mineral** *naturally solid material with distinctive internal crystal structure*

**mitochondria** *cell organelles (descended from respiring bacteria) that use oxygen to break up food molecules and create energy*

**mitosis** *cell division by fission or budding*

**motile** *able to move about*

**mutant** *a member of a population carrying one or more new genes*

**MYA** *million years ago (also, MYA)*

**mycorrhizal** *type of fungus usually living among plant roots*

**Neanderthal** *form of* Homo sapiens *living from 100,000 to 30,000 years ago who were not direct ancestors to human*

**nebula (plural: nebulae)** *tenuous cloud of gas and dust in interstellar space*

**nematocyst** *cellular organelle common in polyps; may be adopted by their predators; a sac containing a long coiled tube with poison "harpoon" tip ejected by water pressure to trap prey*

**neoteny** *retention of some juvenile characteristics in adulthood*

**neutron** *chargeless heavy subatomic particle that appears along with protons in all nuclei other than hydrogen*

**nitrogen fixation** *a process that causes free nitrogen to combine chemically with other elements to form reactive compounds*

**notochord** *flexible rodlike longitudinal structural element in primitive chordates*

**nucleic acid** *compounds that direct the course of protein synthesis, thereby regulating cell activity*

**nucleotides** *building blocks of nucleic acids, composed of nitrogen, sugar, and phosphate group*

216

**nucleus (of atom)** *small, heavy central portion of an atom, composed of protons and neutrons*

**nucleus (of cell)** *central cell organelle of eukaryote, containing DNA and protein within a membrane*

**organelle** *any of a number of organlike structures in eukaryotic cells*

**organic compound** *substance that contains carbon*

**oxidize** *to combine with oxygen*

**ozone** *molecule of three oxygen atoms; blocks UV light in upper atmosphere*

**Pangaea** *hypothesized protocontinent that, several million years ago, broke up into precursors of present continents*

**paramecium** *protist commonly found in pond water; slipper-shaped with coordinated parallel lines of rowing cilia*

**parthenogenesis** *development of a new individual from an unfertilized sex cell, occurring among lower plants and invertebrate animals*

**photoautotroph** *life-form living off light and carbon dioxide*

**photoheterotroph** *life-form living off light and complex compounds*

**photon** *particle of light*

**photosynthesis** *synthesis of chemical compounds with the use of light*

**phylum** *taxonomic group; see* family

**physics** *a science that deals with the structure of matter and the interactions between fundamental constituents*

**physiology** *metabolic processes of an organism*

**planetesimal** *small solid celestial bodies at an early stage of solar system*

**plankton** *minute animal, plant, and protist life of a body of water, especially on its surface*

**plant** *multicellular eukaryotes containing plastids, developing from nonblastular embryos*

**planula** *offspring of medusae, the second free-swimming stage of a polyp's life cycle*

**plasma (electrical)** *a collection of charge particles containing equal numbers of positive ions and electrons*

**plastid** *photosynthetic organelles in eukaryotes; descended from photosynthesizing bacteria*

**plate tectonics** *movement of the crustal plates comprising Earth's surface, with continents "riding" them*

**polyp** *simple animal classed as a cephalopod and attached to a seafloor substrate; e.g. anemones, corals*

**predacious** *living by preying on other animals*

**primary producer** *an organism that converts energy from the Sun or from inorganic substances to produce organic compounds*

**primates** *mammals including lemurs, monkeys, apes, and man that have features including grasping hands, flattened nails, relatively large brain, etc.*

**Prokaryota** *superkingdom comprising single kingdom: Monera*

**prokaryote** *(pronounced pro-CARRY-oat) bacterium, also called moneran, without nucleus*

**propagule** *a sexual or asexual reproductive particle, such as a gamete or spore*

**proteins** *large molecules made of chains of amino acids; basic building blocks of life-forms*

**protist** *small protoctist*

**protoctist** *eukaryotic unicellular organisms; all living things other than plants, animals, fungi, and bacteria (lit: "first being")*

**proton** *a stable, positively charged subatomic particle that is a constituent of the nucleus of every atom*

**proton motor** *micromotor, powered by proton motive force, that spins locomoting flagella of some cells*

**protozoan** *eukaryotic heterotroph microbe, usually motile at some stage*

**quantum mechanics** *physics of atomic and subatomic systems and their interaction with radiation*

217

**quantum chromodynamics** *the theory that describes the action of strong nuclear forces, stronger than electrical forces*

**quantum electrodynamics** *the theory of the interaction of charged particles with the electromagnetic field*

**quark** *fundamental constituent of protons and neutrons, in much the same manner that the latter are constituents of nuclei*

**radiation, adaptive** *divergence of multiple species from single ancestral lineage radicals, free highly reactive molecules containing at least one unpaired electron*

**reptiles** *class of air-breathing, generally scaly, vertebrates that includes snakes, lizards, alligators, and dinosaurs*

**respire** *to inhale and exhale air, taking up oxygen and expelling carbon dioxide through oxidation*

**RNA** *ribonucleic acid; related to DNA but single-stranded; several varieties play roles in copying DNA and synthesizing proteins*

**sex** *any process recombining genes from more than a single source to form a single being*

**species (singular and plural)** *group of organisms capable of interbreeding*

**spirochete** *slender, spirally undulating bacterium*

**spore** *unicellular environmentally resistant dormant or reproductive body produced by plants and some microorganisms*

**Stone Age** *period dating from first use by humans of stone tools 2,500,000 years ago*

**stromatolite** *layered sedimentary rock formations bound by colonial bacteria*

**subduction zone** *region where one tectonic plate pushes downward beneath the edge of another into Earth's upper mantle*

**symbiogenesis** *evolution of new being from mergers of independent organisms*

**symbiont** *any organism involved in an intimate and protracted association with another organism of a different species*

**symbiosis** *intimate association over time of two dissimilar organisms*

**synergy** *working together*

**tectonics** *see* plate tectonics

**tensegrity** *Buckminster Fuller term for rigid systems of combined compressive and tensional integrity; exhibited in biology by internal certain cellular architectures and bone-muscle-tendon systems*

**test** *loose-fitting shell secreted by some protists, such as diatoms*

**tropopause** *upper boundary of troposphere, below stratosphere*

**troposphere** *five- to ten-mile blanket of air closest to Earth's surface wherein temperature decreases with distance from surface*

**ultraviolet light** *electromagnetic radiation having wavelengths slightly shorter than those of visible light*

**unicellular** *single-celled*

**universe** *all matter and space containing some million million galaxies*

**UV (see *ultraviolet*)** *abbreviation for ultraviolet*

**vacuole** *small cavity-shaped cell organelle performing functions such as storage, ingestion, digestion, excretion, and the like*

**visible light** *electromagnetic radiation of wavelengths ranging around half a thousandth of a milimeter*

**x-radiation** *electromagnetic radiation characterized by wavelength 10 to 100,000 times shorter than that of visible light*

## ACKNOWLEDGMENTS FOR THE
### *WALK THROUGH TIME* EXHIBITION AND BOOK

The original one-mile "A Walk through Time . . . from stardust to us" exhibit was created by employees of the Hewlett-Packard Company with central contributions by outside experts under the Company's sponsorship. It was first presented as a context-setter at HP Laboratories' Celebration of Creativity on Earth Day, 22 April 1997, at HP Laboratories sites in Palo Alto, California; Bristol, England; and Tokyo, Japan. Sidney Liebes is responsible for the original *Walk* concept. The project was sponsored by Joel Birnbaum, HP Senior Vice President for Research and Development, and Director of HP Labs. The project was co-led by him, Laurie Mittelstadt, and Barbara Waugh. Lois Brynes, of Deep Time Associates, heroically conducted the "Walk" exhibit content research, photo selection, and text composition, with HP volunteer copy editor assistance, all in an impossibly short six weeks.

Many others, most but not all of whom were Hewlett-Packard employees, made exceptional contributions to the Celebration and Walk. They include Geoff Ainscow, Cindy Alfieiri, Wally Austin, Brian Barry, Lynn Beckley, Betty Belloli, Jean-François Berche, Sally Cohn Berche, Steve Bicker, Gail Burke, Jackie Burleigh, Aleta Chandra, Cy Class, Maria Colin, Peter Cook, Linda Creatore, Wayne Davies, Chris de Vos, Jasmale Dhillon, Don Dunphy, Dennis Freeze, Nancy Freeze, Toshio Fujiwara, Joan Gallicano, Henrietta Gamino, Bill Gibson, Ruth Gilombardo, Caitlin Hall, Lorene Hall, Sharon Hanrahan, Ian Hardcastle, Tom Hornak, Pat Ichelson, Tak Kamae, Ed Karrer, Barbara Keen, Rhonda Kirk, Michi Kitaura, Cristina Konjevich, Joe Kral, Dick Lampman, Rob Lawrence, Debbie Leos, Kay Lichtenwalter, Linda Liebes, Andrew Liu, Robin Locklin, Rhonda Louie, Loretta Lovingood, Frank Lucia, Rich Marconi, Gina Massey-Parnis, Shirley McFadden, Molly Megraw, Raakhee Mistry, Laurie Mittelstadt, Jim Nunes. Also HP Labs Facilities staff: Hans Obermaier, Thorsten Obermaier, Ian Osborne, John Otsuki, Gina Massey-Parnis, Chandra Patel, Herlinda Perez, Rick Pierce, Lew Platt, Jim Raddatz, Johnny Ratcliff, Cheryl Ritchie, Mike Rodriquez, Penny Rose, Bina Shah, Ruth Shavel, Jim Sheats, Bill Shreve, Darlene Solomon, Roger Sperry, Roxann Stephens, Randy Strickfadden, Srinivas Sukumar, Chie Tajima, John Taylor, Jean Tully, Dick Van Gelder, Chuck Untulis, Shalini Venkatesh, Jim Wack, Rick Walker, Hazen Witemeyer, Art Wittke, Sheila Worland, Rosanne Wyleczuk. Additionally, George Radominski of HP Corvallis, OR, Karl Tiefert of HP Components Group, CA, and Joe Thomas of HP Greeley, CO, played outstanding championship roles in bringing the "Walk" to their sites and cities.

Appreciation is expressed to the 70 artists, photographers, and copyright owners whose contributions are credited elsewhere in this volume, for generous permission to use their illustrations. The original exhibit panels were designed by Lisa Otsuki, of SOZO, Santa Clara, CA. Exhibition panels were printed on Hewlett-Packard DesignJet large-format printers by John Otsuki of Capital Color, a division of Deltrek Inc., Santa Clara, CA, and by Mike Pittenger of Focus Digital Services, Nashville, TN. Special thanks go to Eva Priplatova of Aventinum Publishing, Czech Republic, for her selfless contribution at the formative stages of the project, resulting in use of the art of Zdenek Burian and Jan Sovak.

The Hewlett-Packard Company, after generously investing resources to create the "Walk" for internal and external presentations, has gifted it to the nonprofit Foundation for Global Community (FGC), Palo Alto, California. FGC is committed to developing and sharing the "Walk" and derivative products throughout the world.

Biologist Lynn Margulis contributed significantly to the "Walk" exhibit. Her book *What Is Life*, co-authored with Dorion Sagan, was a prime reference for the exhibit text. Other scientists contributing advice and support included A. Anbar, Mike Bolte, A. Brack, P. Braterman, S. Chang, K. Cowing, D. Cruikshanik, Bertha Cutress, W. Davis, Laurie Godfried, J. Grotzinger, S. Liebman, Jere H. Lipps, Marc McMenamin, B. Minard, Norman Pace, Y. Pendleton, Hans Reichenbach, A. Rogerson, R. Rye, J. William Schopf, E. Shoch, John Sieburth, Mitchell Sogin, Michael Soule, Rich Wikander, and E.O. Wilson. Full responsibility for deficiencies in the "Walk" exhibit text is assumed solely by Sid Liebes.

Special appreciation is extended to book packagers Philip and Manuela Dunn of The Book Laboratory, Inc., Mill Valley, CA, for championing and formatting this book, in collaboration with Renée Harcourt, Janet Mumford, and Ali Meyer of i4 Design. Emily Loose, senior editor at John Wiley & Sons, Inc., helped guide and edit this book.

FGC acknowledges legal advice of Robert F. Levine, of Levine, Thall, Plotkin & Menin, New York, NY, and Terence M. Kelly, of Ritchey, Fisher, Whitman & Klein, Palo Alto, CA; "Walk" Logo design by Ian Smith; and QuarkXPress exhibit panel editing by George Lau.

Geoff Ainscow, Wileta Burch, Bill Devincenzie, Lisa Friedman, Joe Kresse, Sid Liebes, Eileen Rinde, Richard Rathbun, and Samantha Schoenfeld, of FGC guide the "Walk" project. Dozens of additional FGC volunteers and friends have devoted hundreds of hours to the project

Co-author Brian Swimme acknowledges for crucial support his gratitude to associate Bruce Bochte, Foundation for Global Community colleagues Wileta Burch, Jim Burch, Sid Liebes, Karen Harwell, and Richard Rathbun. Co-author Elisabet Sahtouris thanks Howard Bloom, James W. Brown, Forrest Iarwain, and Don Ingber. Sid Liebes expresses appreciation to family members Linda, Karen, David, Marjorie and Helen for support and understanding of the impact of the "Walk" project on family life these past two years. The role of his father in opening Sid's eyes to the wonders of the wilderness was fundamental to Sid's commitment to the "Walk" project.

Funding support for the "Walk" project has been received from the Hewlett-Packard Company, the Richard & Rhoda Goldman Fund, the Compton Foundation, and The Foundation for Global Community.

# ILLUSTRATION ACKNOWLEDGMENTS

**Carmen Aguilar-Diaz:** 8 (right), 37 (above). Adapted from graphic by **Jesse Anderson,** courtesy **New England Science Center:** 45. **Stanley Awramik:** 76, 77 (right). **Elso Barghoorn:** 62. **Nic Bishop,** courtesy **New England Science Center:** 166 (right). **Lucian Bordeleau:** 150. **Lois Brynes:** 19, 22, 81, 95, 99, 110 (left), 134, 138, 146 (right), 147 (above right), 154, 163, 187, 196, 201 (lower right). **Zdeněk Burian** © **Jiri Hochman** and **Martin Hochman:** 129, 153, 157, 159, 165 (above), 173, 197, 199, 200-201 (with exception lower right). **Fred Campbell:** 51. **Susan Campbell:** 97. **David Caron,** Woods Hole Oceanographic Institution: 88, 123 (all). **Esmeralda Caus:** 122. **David Chase:** 60. **Preston Cloud** © **Janice Cloud:** 69. **Charles E. Cutress, Jr.** © **Bertha M. Cutress:** 121 (above). **Kathryn Delisle:** 100. **M. Despezio,** courtesy Marine Biological Laboratory, Woods Hole: 72. **Brian Duval:** 112, 113 (above and below); with **Lynn Margulis:** 116-117. Maps reproduced with permission from Britannica CD 97 © 1997 by **Encyclopaedia Britannica,** Inc.—-adapted from **C.R. Scotese,** University of Texas, Arlington: 124-125. **Richard Ettinger:** 70-71. **Daphne Fautin:** 109 (below). **Lynda Goff:** 104. **Stjepko Golubic:** 65, 195 (right). **Elisabeth Gong-Collins:** 42 (left). **Ricardo Guerrero:** 35 (below). **Johannes H. P. Hackstein:** 83. **William K. Hartmann,** from *The History of the Earth* by William K. Hartmann and Ron Miller, © 1991: 8 (left), 11 (below), 12, 13 (left), 188 (above left and right; below right). **Patrick Holligan:** 171 (right). **JASON Foundation for Education:** 26 (left), 135, 141. **Ludwig Kies:** 85, 93. **Annelies Kleÿne:** 170-171. **Andrew Knoll:** 110 (right). **Alan Kuzirian,** courtesy Marine Biological Laboratory, Woods Hole: 165 (bottom left). **Jean Cyril Lecomte, C.N.R.S.:** 133 (below right). **John Lee:** 109 (above). **Roger Leo,** courtesy New England Science Center: 146 (left). **Annabel Lopez** © **Ricardo Amils:** 26 (right). **Christie Lyons:** 35 (above), 57, 86-87, 101, 118-119, 139, 143, 147 (below). **Lynn Margulis:** 9 (right) 33, 42 (right), 46-47, 50, 54, 55, 61, 63 (above), 77 (center), 151 (of specimen discovered by **Mark A. McMenamin**), 176; with **Brian Duval:** 116-117. **Marine Biological Laboratory, Woods Hole:** 120 (below), 121 (below right), 133 (above, and below left), 165 (below right). **Mark A. McMenamin:** 117 (right). **Norman Meinkoth:** 121 (below left). **Ron Miller,** from *The History of the Earth* by William K. Hartmann and Ron Miller, © 1991: 13 (right), 188 (below left and right; above right), 189 (below left). **NASA:** 10, 11 (above), 17, 25, 59, 144-145, 181. **National Center for Atmospheric Research:** 20, 49, 75, 107. **Ocean Drilling Program:** 169. **Patrick O'Connor,** courtesy New England Science Center: 158. **Jerome Paulin** © **Mrs. Jerome Paulin:** 179. **Norbert Pfennig:** 34-35, 42. **Jeremy Pickett-Heaps:** 98, 175 (below). **Hans Reichenbach:** 43. **R. George Rowland,** courtesy Marine Biological Laboratory, Woods Hole: 9 (left), 175 (above). **Royal British Columbia Museum:** 132, 193. **Peter Sawyer,** courtesy Smithsonian Institution: 31. **J. William Schopf:** 39 (above). **Stephen and Silvya Sharnoff:** 147 (above left). **John Sieburth:** 105. **Jan Sovak:** 131, 166 (left), 177, 178, 185, 190, 194-195. **Richard Stemberger:** 9 (middle), 120 (above). **John Stolz:** 37 (below). **Paul Strother:** 63 (below). **Gonzalo Vidal:** 77 (left), 111 (left and right), 117. **Peter Westbroek:** 39 (below). **Donald Zinn** © **Margery P. Zinn:** 130.

220

# INDEX

221

223